Testing, Rationality, and Progress

**The Worldly Philosophy:
Studies at the Intersection of Philosophy
and Economics**

General Editor: Philip Mirowski, University of Notre Dame

*Testing, Rationality, and Progress: Essays on the Popperian Tradition
in Economic Methodology*
   by D. Wade Hands, University of Puget Sound

Edgeworth on Chance, Economic Hazard, and Statistics
   edited by Philip Mirowski, University of Notre Dame

Testing, Rationality, and Progress

Essays on the Popperian Tradition in Economic Methodology

D. Wade Hands

Rowman & Littlefield Publishers, Inc.

ROWMAN & LITTLEFIELD PUBLISHERS, INC.

Published in the United States of America
by Rowman & Littlefield Publishers, Inc.
4720 Boston Way, Lanham, Maryland 20706

British Cataloging in Publication Information Available

Library of Congress Cataloging-in-Publication Data

Hands, D. Wade
Testing, rationality, and progress : essays on the Popperian
tradition in economic methodology / D. Wade Hands.
p.   cm. — (Worldly philosophy)
Includes bibliographical references (p.   ) and index.
1. Economics—Methodology. I. Title. II. Series.
HB131.H36   1992   330'.072—dc20   92–8568   CIP

ISBN 0-8476-7724-9 (cloth : alk. paper)

Printed in the United States of America

The paper used in this publication meets the minimum requirements of
American National Standard for Information Sciences—Permanence of
Paper for Printed Library Materials, ANSI Z39.48–1984.

To My Early Teachers in These Matters:
Scott Gordon and Noretta Koertge

Contents

Preface

Economic methodology has become a minor growth industry during the past decade. Much of this growth has, in one way or another, been related to the philosophy of Karl Popper. More than any other school of philosophy, the Popperian tradition (defined broadly enough to include those as disparate as Lakatos and Bartley) has represented the dominant voice in contemporary methodological discourse.

The following chapters provide my evaluation of the Popperian tradition in economic methodology. The first ten chapters are all previously published papers (and they appear roughly in chronological order). Some of these chapters were originally published as journal articles; some were published as long reviews; while still others originally appeared in conference volumes. All of the previously published papers have been rewritten stylistically, and hopefully improved, but I have sincerely tried to leave the content of these papers as it originally appeared. The first ten chapters thus represent a rather extensive narrative regarding the Popperian tradition in economic methodology, a narrative from the viewpoint of one particular participant-observer: me.

Chapter 11 is new and previously unpublished. It was written as the concluding chapter to this volume, and it represents my current evaluation of the Popperian tradition in economic methodology. The central thesis of this final chapter is that, while it *is* possible to defend the Popperian tradition, much of the defense still needs to be articulated; and once articulated, the finished product may be rather weak-kneed from the viewpoint of many "Popperian" economic methodologists.

Different readers of Chapter 11 will no doubt have quite different reactions to it. A few readers may find my conclusions to be a complete abandonment of the Popperian philosophical tradition—a surrender to the recent cavalcade of perspectival and contextual voices. Certain other readers may respond in the opposite way, finding my conclusions to be (still) much too traditional, much too concerned with the epistemological privileging of certain modes of discourse at the expense of others. Such disparate interpretations are probably inevitable, and I will only be concerned if the overwhelming response of the methodological community is clearly in favor of *one* of these interpretations over the other. As long as the criticisms on these two sides remain evenly balanced, I will not be particularly troubled.

Many, many people have commented on and contributed to this work in many, many ways. It is certainly not possible to name everyone who has offered a useful opinion or otherwise influenced the ultimate shape of this work. Nonetheless, I will try to name some of the individuals who have been influential. With apologies to those I am neglecting, I would like to mention Roger Backhouse, Jack Birner, Mark Blaug, Larry Boland, Bill Brown, Bruce Caldwell, Bob Coats, Avi Cohen, Neil de Marchi, Craufurd Goodwin, Scott Gordon, Bert Hamminga, Dan Hammond, Dan Hausman, Visa Heinonen, Abe Hirsch, Kevin Hoover, Terence Hutchison, Arjo Klamer, J. J. Klant, Christian Knudsen, Noretta Koertge, Maurice Lagueux, Richard Langlois, Donald McCloskey, Uskali Mäki, Phil Mirowski, Mary Morgan, Robert Nadeau, Alan Nelson, Bart Nooteboom, John Pheby, Jukka-Pekka Piimies, Alex Rosenberg, Malcolm Rutherford, Andrea Salanti, Warren Samuels, Jorma Sappinen, Margaret Schabas, Jeremy Shearmur, Mike Veseth, Roy Weintraub, and Nancy Wulwick.

I would also like to thank the University of Puget Sound for providing an environment where this type of interdisciplinary work is not only acceptable, but frequently encouraged and often rewarded. The work has been helped along by two Martin Nelson Summer Research Grants, as well as a number of smaller travel and research grants administered through the University Enrichment Committee. Despite my general accommodation to the folkways of modern academic life, the one thing I have steadfastly resisted is the "do your own word processing on your own PC" revolution. For this reason, I would like to give particular thanks to the people who were ultimately responsible for making this work available to you: Reggie Tison, Judith Sears, and Laura Kiely at the University of Puget Sound, and a number of anonymous but appreciated members of the Notre Dame Typing Pool. All of their work, along with that of my copy-editor Pat Merrill, has been greatly appreciated. Finally, I would like to thank my editor Jonathan Sisk and all of the people at Rowman and Littlefield who have contributed so diligently to the preparation of the book you have before you. To all of these people, as well as my teachers and my students, I say thanks.

Acknowledgments

Chapters 1 and 3 originally appeared in *Philosophy of the Social Sciences* and are reprinted with permission.

Chapters 2 and 4 originally appeared in *History of Political Economy* and are reprinted with permission from Duke University Press.

Chapter 5 originally appeared in *Review of Political Economy* and is reprinted with permission from Edward Arnold Publishing.

Chapters 6 and 7 are both reprinted with permission from Cambridge University Press. Chapter 6 originally appeared in *Economics and Philosophy*. Chapter 7 is reprinted from *The Popperian Legacy in Economics*, Neil de Marchi, ed., 1988.

Chapter 8 is reprinted with permission from Kluwer Academic Publishers. It originally appeared in *Post-Popperian Methodology of Economics: Recovering Practice*, Neil de Marchi, ed., 1992.

Chapter 9 originally appeared in *Appraising Economic Theories: Studies in the Methodology of Scientific Research Programmes*, Mark Blaug and Neil de Marchi, eds., 1991, and is reprinted with permission of Edward Elgar Publishing.

Chapter 10 originally appeared in *Finnish Economic Papers* and is reprinted with permission.

Chapter 11 was written for this volume and not previously published. Portions of the chapter have been presented in various forums, including the FPPE Workshop on Economic Methodology in Mekrijärvi, Finland, the philosophy department of the University of Quebec at Montreal, and at the University of Notre Dame.

1

The Methodology of
Economic Research Programs

This paper (Hands, 1979) originally appeared in Philosophy of the So-cial Sciences *as part of a review symposium on Latsis,* Method and Appraisal in Economics *(1976a). The other papers appearing in the symposium were Agassi (1979) and Archibald (1979).*

The essay was written while I was still a graduate student. At the time I was toying with the idea of writing my doctoral dissertation on the topic of Lakatos and economics. Although I ultimately wrote on an unrelated topic, this paper is an accurate reflection of my initial eval-uation of Lakatos's work and its relationship to the Popperian tradi-tion.

The paper is generally optimistic about future work on Lakatos and economics. For this reason, it makes an ideal first chapter to the cur-rent volume. In the mid-1970s, many people writing on economic methodology were optimistic about Lakatos, and this essay reflects that hopeful spirit.

Regarding the details of the paper, it focuses specifically on the "conventionalist" element in the Popperian tradition. Economists have generally missed this conventionalist theme in the philosophical literature, and I argued that it has led them to a rather naive view of both Popper and Lakatos. This conventionalism is an aspect of the Popperian tradition that is seldom reemphasized in later chapters. What later chapters do share with this discussion is the general position that economists should be more careful in their reading of the philosophical literature they use.

Economists have recently been drawn into the controversy surrounding the late Imre Lakatos's work in the history and philosophy of science.

1

The volume examined here[1] has been followed by a number of articles on the topic,[2] and it seems very likely that the trend will continue. Work in this area has been animated by the economics profession as well as by those writing on the history and philosophy of science more generally.

The rise of "growth of knowledge" theories[3] and the recent resuscitation of two-way communication between historians and philosophers of science has helped to broaden the metascientific discussion about economics. From the economic side there has been a general revival of interest in methodological issues. This revival has been initiated in part by theoretical controversy within mainstream economics, controversy caused by the breakdown of the previously dominant neoclassical synthesis.[4] Alternative scientific methodologies provide economists with alternative vehicles for appraising the history of their discipline and offer the potential for a methodological reconciliation of current theoretical controversies. For philosophy of science, economics provides both a "testing ground" and an opportunity to explore scientific "rationality" in a context different from the traditional domain of natural science.

Spiro Latsis's *Method and Appraisal in Economics* (1976a) consists of a number of papers, most by economists, which grew out of the Nafplion Colloquium on Research Programmes in Physics and Economics held in Nafplion, Greece, in September 1974. The papers in the volume focus on a number of different issues: some of them offer general criticism of the application of Lakatosian methodology, some offer general praise, and still others provide specific case studies in the history of economic thought. The volume as a whole is excellent and it will undoubtedly serve as an important point of departure for future work in this area.

Since there are many summaries available in the literature I will not attempt to summarize the central thesis of Lakatos's methodology of scientific research programs (MSRP).[5] Nor will I summarize each of the various contributions to the Latsis volume or undertake a detailed criticism of any individual essay. Instead, I will offer a broad criticism that applies to almost all of the papers. The criticism is that the papers in the volume neglect an important part of Lakatos's work that needs to be emphasized in order properly to "apply" the MSRP to economics. I will argue that this aspect of MSRP is a result of its Popperian antecedency and that failure to underscore this aspect permits a perspective on the relationship between the history of science and the methodology of science that is inconsistent with Lakatos's central argument. In other words, Lakatos's position has some important implications that are either disregarded or not fully understood by the majority of the authors in the Latsis volume. This leads to inconsistencies and it limits the value of these papers in a general appraisal of the MSRP.

The result of this lacuna is that supporters of Lakatosian methodology are often found to advocate very non-Lakatosian views on the relationship between history and philosophy of science. On the other hand, we will find that critics of Lakatos who advocate a more "Popperian" methodology are often skeptical of Lakatos in precisely the areas where his work is a refinement of Popper's.

The discussion proceeds by pointing out an important "conventionalist"[6] aspect in Karl Popper's philosophy, and then demonstrating that the MSRP is also "Popperian" in this conventionalist sense. It will then be shown that this aspect of Popperian philosophy is fundamental—not adventitious—to the appraisal of the Lakatosian program. Finally, it will be argued that Lakatos's view has certain characteristics related to this conventionalist component that are not recognized by the majority of the contributors to *Method and Appraisal*.

THE RISE OF DOGMATIC FALSIFICATIONISM

As Lakatos makes very clear, the dominant tradition in the philosophy of science prior to the early part of the twentieth century was foundationalist: "knowledge meant proven knowledge" (1970, p. 91). Good scientific method was a way of extending that proven knowledge to new domains. Empirical justificationists argued that acceptable scientific knowledge must be "proven": either proven "facts" or statements logically derived from these proven "factual propositions." According to Lakatos, classical empiricists accepted as axioms only a relatively small set of factual propositions, which expressed the "hard facts." The truth value of these facts was established by direct experience and they constituted the empirical basis of science (Lakatos, 1970, p. 94).

Since deductive logic allows only nonampliative inferences, it was necessary to develop an inductive logic that would transmit truth from these factual propositions to "general laws." In this way the scope of proven knowledge could be extended from the empirical basis to general laws by truth-preserving ampliative inferences.[7] Difficulties associated with the Humean problem of induction eventually led philosophers to abandon the belief that such an inductive logic could be found.

One attempt to reconcile the empirical basis of proven knowledge with the problem of induction was dogmatic or naturalistic falsificationism. The falsificationist argument was that, while science could not prove, it could conclusively disprove. Armed with the logic of modus tollens and the given empirical basis, dogmatic falsificationism set out to extend the body of proven knowledge by a process of elimination. Theories are fallible, and once they fail, they cease to be contenders for knowledge. Lakatos characterizes dogmatic falsificationism in the following way:

> Dogmatic falsificationism admits the fallibility of all scientific theories without qualification, but it retains a sort of infallible empirical basis. It is strictly empiricist without being inductivist; it denies that the certainty of the empirical basis can be transmitted to theories. (Lakatos, 1970, p. 96)

> According to the logic of dogmatic falsificationism, science grows by repeated overthrow of theories with the help of hard facts. (Lakatos, 1970, p. 97)

Lakatos presents at least two criticisms of this dogmatic falsificationist methodology. The first is a problem that stems from its foundation in naive empiricism; this is one of the problems that led Popper to develop his more sophisticated version of falsificationism. The second conundrum is sometimes referred to as the "Duhemian problem" and it applies to all forms of falsificationism, dogmatic as well as more sophisticated versions.[8] In some respects, it was this second problem—the Duhemian problem—that led Lakatos to abandon falsificationism altogether and develop his MSRP. The Duhemian problem will be neglected here, though, since it is more important to falsificationist methodologies than to Lakatos's MSRP.[9] On the other hand, the first problem—the problem of the empirical basis—does need to be examined for a proper understanding of Lakatos's work and its place in the Popperian tradition.

THE PROBLEM OF THE EMPIRICAL BASIS

According to Lakatos, the most important problem for dogmatic falsificationism—and the one that Popper sought to eliminate in his more sophisticated falsificationism—is the problem of the empirical basis. The difficulty is that "facts" are "theory laden." The process of obtaining knowledge of the empirical basis is a process that involves a contribution of the observer to the observed. Observations are not something we "have," but something we "make" (Popper, 1972, p. 342). Conflicts between a scientific theory and data do not constitute conflicts between a theory and nature, but inconsistency between a scientific theory and an observation theory. There is no guarantee that the observation theory is correct: there is no guarantee that the "facts" correspond to the facts.[10] The point is clearly stated in a recent introductory text on the philosophy of science:

> Descriptions of observations cannot be entirely independent of theory either in form or in content. There are no modes of description which remain invariant under all changes of theory. The way in which observations are described changes where theory changes. The accepted way of explaining phenomena enters into the very meaning of the terms used to describe them. It seems to be generally agreed among philosophers, now, that the ideal of a descriptive vocabulary which is applicable to observations, but which is entirely innocent of theoretical influence, is unrealizable. (Harre, 1972, p. 25)

Lakatos makes a similar argument:

> In particular, for classical empiricists the right mind is a *tabula rasa*, emptied of all original content, freed from all prejudice of theory. But it transpires from the work of Kant and Popper—and from the work of psychologists influenced by them—that such empiricist psychotheraphy can never succeed. For there are and can be no sensations unimpregnated by expectations and therefore there is no natural (i.e., psychological) demarcation between observational and theoretical propositions. (Lakatos, 1970, p. 99)[11]

"CONVENTIONALIST" FALSIFICATIONISM

Popper's "falsificationist" philosophy of science does not rely on the classical empiricist's empirical basis. Instead, Popper proposes a "revolutionary conventionalism" where "observation statements" are accepted only by decision (Lakatos, 1970, p. 106). The decision to accept certain "basic statements" as potential falsifiers is a result of Popper's position that all theories are fallible and that there are fallible theories involved in "observation." These "basic statements" are accepted by "convention" as part of unproblematic background knowledge for the purpose of testing the theory at hand; "he may call these theories . . . 'observational': but this is only a manner of speech which he inherited from naturalistic falsificationism" (Lakatos, 1970, pp. 106-7). Popper thus substitutes an "empirical basis" for the empiricist's empirical basis. Popper's own statement of this is as follows:

> In introducing the term "empirical basis" my intention was, partly, to give

an ironical emphasis to my thesis that the empirical basis of our theories is far
from firm; that it should be compared to a swamp rather than solid ground.

Empiricists usually believed that the empirical basis consisted of absolutely
"given" perceptions or observations, of "data," and that science could build on
these data as if on rock. In opposition, I pointed out that the apparent "data" of
experience were always interpretations in the light of theories, and therefore
affected by the hypothetical or conjectural character of all theories.

... [T]he process of interpretation is at least partly physiological, so that there
are never any uninterpreted data experienced by us: the existence of these
uninterpreted "data" is therefore a theory, not a fact of experience, and least of
all an ultimate or "basic" fact.

Thus there is no uninterpreted empirical basis; and the statements which
form the empirical basis cannot be statements expressing uninterpreted "data"
(since no such data exist) but are, simply, statements which state observable
simple facts about our physical environment. They are, of course, facts
interpreted in the light of theories: they are soaked in theory, as it were. (Popper,
1965, p. 387).[12]

Thus it is clear that Popper does not advocate dogmatic falsificationism, that he "separates rejection and disproof, which the dogmatic
falsificationist had conflated" (Lakatos, 1970, p. 109). The decision to
"falsify" is a "decision"—based on the simple rule that, when a the-
ory-laden observation is inconsistent with a scientific theory, one must
relegate the observation to unproblematic background knowledge and
reject the scientific theory.

Popper's simple decision rule is not present in his writings on "sophisticated" falsificationism, but only in its "naive" form. In the sophisticated form Popper allows a theory to be protected from falsification by a single conflict between the theory and a contrary "observation."
But this sophisticated falsificationism does not eliminate the conventional element in the appraisal of scientific theory; it only reduces it.
We cannot avoid the decision as to which propositions should be the
"observational" and which the "theoretical" ones.

It is standard practice in elementary presentations of Popper's philosophy of science to stress his concern for the method of falsification
rather than confirmation—his "solution" to the problem of induction—
and not pursue his notion of the empirical basis in any depth. Such
presentations lead one to believe that Popper's work is adequately
represented by what Lakatos (1970, p. 181) calls $Popper_0$. $Popper_0$ is
"the dogmatic falsificationist who never published a word," as contrasted to $Popper_1$ the naive falsificationist and "$Popper_2$ the sophisticated falsificationist." Lakatos argues that "the real Popper consists of
$Popper_1$, together with some elements of $Popper_2$," (1970, p. 181). The
issue of concern here is not the relatively ambiguous demarcation of

Popper₁ and Popper₂. The issue is that Karl Popper is unambiguously not a dogmatic falsificationist, and never has been.

Lakatos's MSRP is a scion of Popper's sophisticated falsificationism. It rests on conventionalist decisions regarding the "empirical basis" just as Popper's method does: "As it stands, like Popper's methodological falsificationism, it represents a very radical version of conventionalism" (Lakatos, 1971a, p. 101). In the next section I will examine the treatment of this conventionalist element in Popper's work and Lakatos's MSRP as it is characterized by the contributors to *Method and Appraisal*.

CONVENTION AND APPRAISAL

Most of the contributors to *Method and Appraisal* fail to emphasize this conventionalist aspect of Popper's and Lakatos's work.[13] In explaining Popper's work, Mark Blaug states,

> He repudiated the Vienna Circle's principle of verifiability and replaced it by the principle of falsifiability as the universal, a priori test of a genuinely scientific hypothesis. The shift of emphasis from verification to falsification is not as innocent as appears at first glance, involving as it does fundamental asymmetry between proof and disproof. (Blaug, 1976a, p. 151)

In a similar way, Terence Hutchison and John R. Hicks [both] point to the problems associated with applying Popper's naturalist falsificationism to economics, where empirical data are not available (implying that there is no problem of the empirical basis in physics);[14] similarly Latsis (1976b, p. 14) contrasts conventionalism and Popperian falsificationism without mentioning the conventionalist element in Popper's work.

It is worth noting that economists' dogmatic characterization is not restricted to those critical of Popper's methodology. If it were restricted to critics, the argument could be made that Popper₀ was only a straw man Popper constructed solely for ease of reprobation, but this is decidedly not the case. Those advocating a strict "Popperian" methodology in economics, such as Hutchison, argue for the position of dogmatic falsificationism. Hutchison (1976, p. 203) mentions Popper's "fallibilism"; but it is the fallibility of scientific theories (and therefore the need for a falsificationist rather than confirmationist methodology) to which Hutchison is alluding—not the fallibility of factual propositions as argued in naive or sophisticated falsificationism.

Throughout *Method and Appraisal*, falsificationism, usually attributed

to Popper, is characterized as a conflict between hard facts and con-
jectured theory. This view not only presents Popper and Lakatos as
more philosophically naive than is actually legitimate; it also leads to
misinterpretations of the relationship between the history of economics and the MSRP.[15]

ECONOMIC THEORY AND THE METHODOLOGY OF SCIENTIFIC RESEARCH PROGRAMS

Lakatos's conventionalism regarding the choice of the empirical basis
has direct implications on his scientific historiography. According to
Lakatos, a general methodology of science stands in roughly the same
relationship to the actual history of science as a scientific theory stands
to empirical facts in its domain. As theory influences the "empirical
basis" decision for the scientist, methodology influences the "empirical basis" decision for the historian of science. "History without some
theoretical 'bias' is impossible" (Lakatos, 1971a, p. 107). The "history of science is a history of events which are selected and interpreted
in a normative way" (Lakatos, 1971a, p. 108).[16] Lakatos makes the
following note on his first historical example:

> The proposition "the Proutian programme was carried through" looks like a
> "factual" proposition. But there are no "factual" propositions: the phrase only
> came into ordinary language from dogmatic empiricism. Scientific "factual"
> propositions are theory-laden: the theories involved are "observation theories."
> Historiographical "factual" propositions are also theory-laden: the theories
> involved are methodological theories. (Lakatos, 1971a, p. 127, n. 60)

Thus it is argued that observations in the history of science are
influenced by methodology in exactly the same way as observations in
science are influenced by scientific theory. The historian of science
selects the "facts" to be considered and the history of science is therefore a "decision" containing the same conventional element as decisions made in scientific practice. Lakatos's argument is a relatively
natural extension of the Popperian conventionalism discussed above.
Since there is interaction between the "facts" of science and the
historian's methodological preferences, conflicts between a scientific
methodology and the history of science cannot constitute a "refutation" of the methodology in any dogmatic falsificationist sense. The
appraisal of alternative scientific methodologies is for Lakatos a much
more complex matter than simply comparing the method to the scientific "facts." Lakatos attempts to overcome this difficulty in elegant

dialectic style, using what he labels a "quasi-empirical meta-criterion."

> I first "refute" falsificationism by "applying" falsificationism (on a normative historiographical meta-level) to itself. Then I shall apply falsificationism also to inductivism and conventionalism, and, indeed, argue that all methodologies are bound to end up "falsified" with the help of this Pyrrhonian machine de guerre. Finally, I shall "apply" not falsificationism but the methodology of scientific research programs (again on a normative historiographical meta-level) to inductivism, conventionalism, falsificationism and to itself, and show that—on this meta-criterion—methodologies can be constructively criticized and compared. This normative-historiographical version of the methodology of scientific research programs supplies a general theory of how to compare rival logics of discovery in which . . . history may be seen as a "test" of its rational reconstructions. (Lakatos, 1971a, p. 109)

It is unclear whether Lakatos is entirely successful in this attempt, or if there is a circularity involved, as Thomas Kuhn (1971, pp. 143-43) has argued. The point is simply that the MSRP was erected on a base composed of certain aspects of Popperian methodology and that acceptance of this foundation is necessary for the acceptance of Lakatos's program. This acceptance also implies a certain perspective on the interanimation of scientific methodologies and the history of science. To claim to be an advocate of Lakatosian methodology and yet hold a view of the history of science that is sycophantic on other earlier methodologies is not only inconsistent; it directs attention away from one of the principal messages of Lakatos's work. If methodologies could simply be "tested" by the history of science in a dogmatic way, then theories could simply be "tested" by observations. While this view is quite common in economic methodology, those who subscribe to it face an inconsistency with Lakatos's work or those aspects of Popperian methodology from whence it sprang.

Failure to accept Lakatos's conventionalism at the meta level implies that his MSRP must also be rejected and, by further implication, that there was never any need to move beyond dogmatic falsificationism. This, of course, does not say that one must accept the Lakatosian framework; it only says that acceptance at the micro level entails its acceptance at the macro level, and vice versa.

Many of the contributors to *Method and Appraisal* appear to be inconsistent in this sense. They cling to the demarcation between the objective (positive) history of economics and methodological prescriptions about how economics "ought" to be done (normative), while advocating the MSRP as the vehicle of appraisal for economic theory. Acceptance of Lakatos's view implies that methodological prescriptions describing what scientists ought to do are inexorably intertwined with a putatively positive presentation of what science has done.

Axel Leijonhufvud (1976, p. 66), for instance, remarks of the "apparent 'drunkard's walk' along and across this sacred line" of normative and positive aspects. He then argues it is unclear whether an instance in the history of science that is inexplicable in the Lakatosian framework should "falsify" the MSRP.

From those aspects of Lakatos's work discussed above, it is apparent that a "drunkard's walk" is an essential element of the MSRP, and that "falsifying" the MSRP represents the elevation of falsificationism to the status of an accepted general theory of rationality. This amounts to answering the Lakatosian question before asking it. The MSRP can be collated to falsificationism, but there must first be a decision regarding the historical "empirical basis" in question. The MSRP must be compared to falsificationism by Lakatos's "quasi-empirical meta-criterion" (1971a, p. 110). To approach it in the direct method of Leijonhufvud amounts to allowing a classical empiricism rejected by Lakatos to enter by the back door. This of course represents an inconsistency for an advocate of the MSRP such as Leijonhufvud.

Blaug, on the other hand, recognizes these implications in Lakatos (Blaug, 1976a, p. 150), but analyzes Kuhn's work in terms of the distinction between "normative methodology" and "positive history" (p. 152)[17] where it is no more applicable than it is to Lakatos. Hutchison argues that a "normative positive confusion" (1976, p. 182) is prevalent in both Kuhn and Lakatos, and yet argues for a "Popperian view" ostensibly free of such confusion. Apparently Hutchison fails to recognize that such "normative positive confusion" is not "confusion" at all but rather an important element of Lakatos's general approach resulting from its Popperian roots.

Overall it seems that the contributors to *Method and Appraisal* have avoided the conventionalist element in both Popper and Lakatos and that this has led them to view the history versus methodology issue in a very nonconventionalist and consequently non-Lakatosian way. This leads to inconsistencies in many of the essays and a failure to grasp the full implications of Lakatos's argument. It seems that these authors are indicative of the economics profession more generally; they are trying to fit contemporary metascience such as the work of Lakatos or Kuhn into a dogmatic empiricism that these philosophers unanimously reject. Contemporary philosophers of science such as Lakatos have developed their work in an effort to explain the growth of knowledge without depending on an unacceptable dogmatic empiricism. In general this does not seem to be recognized by economists.

In good conscience it must be noted that some of the essays—specifically those of A. W. Coats (1976), Neil de Marchi (1976), Spiro Latsis (1976b), and Herbert Simon (1976)—tactfully avoid the problems mentioned above, although they still lack any explicit delineation of the conventionalist element and its implications.

CONCLUSION

It should be noted in closing that the above criticisms are not intended to devalue the general importance of the essays contained in *Method and Appraisal*, but are only meant to elucidate important refinements that must be made if we are to accurately apply the MSRP to economics.

These criticisms suggest that economists who are involved in such work should start with a "deeper" reading of both Popper and Lakatos. The conventionalist influence in the works of both is more than a philosophic subtlety and has substantive implications for the MSRP and economics. The normative/positive distinction long cherished by the economic profession may be applicable in the policy situations where it is usually made, but it is inappropriate for comparing methodologies of science. Used in this way it represents a type of methodological atavism. The above discussion also warns against "applying" a scientific methodology to economics without an extended historical investigation of its relation to other methodologies.

Even with the above criticisms, it appears that work in the methodology of economic research programs has stepped from nonexistence directly into a "progressive" period with the publication of *Method and Appraisal*. The arguments set forth above should be taken as a guide to refinement, not as an indictment for abandonment.

NOTES

1. Latsis (1976a).
2. Blaug (1976b) and O'Brien (1976) for example.
3. Broadly, this covers recent work by authors such as Paul Feyerabend, Thomas Kuhn, Imre Lakatos, and Stephen Toulmin, but it also includes certain aspects of more traditional views within the philosophy of science, such as Karl Popper.

4. The use of the term *revolution* is not intended to bias the discussion, even though an important part of the work discussed is concerned with the question of whether Keynesian economics constitutes a revolution. See Blaug (1976a) and Leijonhufvud (1976).

5. Lakatos (1970) and Lakatos (1971a) are the best sources. With a caveat regarding the issues discussed below, Blaug (1976a), Leijonhufvud (1976), and Latsis (1976b) all present brief and useful summaries of the MSRP.

6. The meaning of "conventionalist" in this context will become apparent as the discussion proceeds.

7. This terminology is due to Salmon (1966, ch. 1).

8. See Laudan (1965) for a brief discussion of the history of the Duheimian problem.

9. While the Duhemian problem is not relevant to the main point of this paper, it remains an interesting issue since it involves the problem of falsification with a ceteris paribus clause—a problem that plays an important role in economic theory.

10. The use of quotation marks separate that which is "observational" by decision as opposed to that which is observational in the sense of classical empiricism follows Lakatos (1970, pp. 98 and 106) and Popper (1965, p. 387).

11. For an example of how far rejection of classical empiricism can go, and for numerous examples of the theory-impregnated nature of scientific data, see Feyerabend (1975a).

12. My apologies for the extreme length of this quote; but considering the extent of disagreement on this point (at least among economists), it seems appropriate.

13. It is immaterial whether this is the result of an imprecise reading of Popper and Lakatos, or whether the distinction was recognized but not considered germane to their arguments. Regardless of the cause, the effect is to prejudice the analysis of Lakatos and the history of economic thought in significant ways. It is the economic "conventional wisdom" to view "falsifying evidence" in a "dogmatic" way, and also to attribute this view to Popper via Terence Hutchison or, incorrectly, Milton Friedman (1953). Thus the burden of proof is on the authors of *Method and Appraisal* to clearly show that they do not follow the conventional wisdom on this issue.

14. Hicks (1976, p. 207) and Hutchison (1976, pp. 181 and 187).

15. Those familiar with Richard G. Lipsey and Peter O. Steiner's introductory text *Economics* (4th ed., 1975) will be amused by Lakatos's choice of example to demonstrate how Popperians advocate a dogmatic falsificationism that Popper himself is too sophisticated to advocate. Lakatos (1971a, pp. 129-30, n. 93) makes reference to William Henry Beveridge's farewell address as director of the London School of Economics, on June 24, 1937, which stresses the influence of the hard facts of the Michelson-Morley experiment on Albert Einstein's subsequent work. Lakatos considers this a distortion of the history of physics by a dogmatic falsificationist. The exact excerpt from Beveridge's speech that Lakatos criticizes adorns the front flyleaf of the

introductory text long considered the most philosophically sophisticated of those used in U.S. universities, that is, Lipsey and Steiner (1975).

16. This statement contains the following note: "Unfortunately there is only one single word in most languages to denote history[1] (the set of historical events) and history[2] (a set of historical propositions). Any history[2] is a theory and value-laden reconstruction of history[1]" (Lakatos, 1971a, p. 128, n. 69).

17. For Kuhn's comments on this interpretation of him by others, see Kuhn (1970c, pp. 233 and 237).

2

The Role of Crucial Counterexamples in the Growth of Economic Knowledge: Two Case Studies in the Recent History of Economic Thought

This chapter, which first appeared as an article (Hands, 1984b) in History of Political Economy, *focuses more on the history of economic theory and less on Popperian philosophy than the other chapters in this volume. It presents two episodes in the history of Walrasian general equilibrium theory where purely mathematical counterexamples initiated the type of theory change, or at least redirection of research, that one would normally associate with a scientific crucial experiment. Although the emphasis is on the recent history of economic thought rather than methodology, the overall message is clearly consistent with the methodological themes in later chapters.*

A "crucial experiment" in science is a simple definitive empirical test that either falsifies or confirms a scientific theory. The ideal crucial experiment excludes all but one of the theories competing as possible explanations. Even a cursory examination reveals that the history of economic science contains few, if any, such crucial experiments.[1] There are a variety of well-known reasons (or, some would say, excuses) for this lack of empirical falsification or attempted falsification within economics. Some of these reasons include the lack of opportunity for laboratory experimentation, the "immunizing stratagem" of ceteris paribus, the general "complexity" of economic phenomena, and the variable nature of the underlying economic system. But regardless of the reasons, the fact remains that if "refutations" or "falsifications" occur at all in economics, they occur as the result of a sustained continuous bombardment of contrary evidence, not as the result of a single strategic crucial experiment.

This absence of empirical crucial experiments prompts some obvious questions for the history and philosophy of economic science. For instance, what is there, within the development of economics, to initiate the type of theory change usually attributed to crucial experiments in natural science? In other words, does economic theory have a methodological surrogate for crucial experiments?

Recently a number of methodologically concerned economists have addressed this question. Craufurd Goodwin (1980), for example, suggests that theory change is the result of "external crucial experiments," that is, changes in the social/political/intellectual environment in which the theory functions. Progress is "associated with major external policy challenges or what society takes to be such challenges" (Goodwin, 1980, p. 618).[2] J. V. Remenyi (1979), on the other hand, argues that empirical anomalies enter the economic core theory principally through the portholes of subdisciplines ("demi-cores") and that change occurs as the result of "core–demi-core interaction."

In certain areas of economic science, these explanations of progressive theory change are quite appropriate. For instance, Goodwin discusses the Keynesian revolution—arguably a case where such external influences had a significant impact. The problem is that cases like the Keynesian revolution do not exhaust theory change in economics. Much progress occurs in a far more pedestrian way essentially insulated from such external influences. What will be discussed below is an alternative, nonempirical and nonexternal, avenue for the theory change in economics.

In some important areas of economic theory (particularly those that have been mathematically formalized), theory change is initiated by mathematical "counterexamples." These counterexamples are not facts in any sense, but yet they provide a way of falsifying certain conjectures about the implications of the theory. The effect of these counterexamples on the evolution of the economic theory is very similar to the effect attributed to an anomalous crucial experiment, that is, the abandonment or at least redirection of an area of research.

The next section below provides two case studies from the relatively recent history of economic theory where such "crucial counterexamples" have significantly affected the development of the theory. Both studies are from Walrasian general equilibrium theory, but similar cases could be cited for international trade theory, growth theory, or any other highly formalized area of economic science.[3] While these case studies have historical interest in their own right, the primary motivation for this study is methodological and these methodological implications are discussed in the final section.

THE STABILITY OF GENERAL EQUILIBRIUM

Leon Walras (1954) only briefly discussed the stability of his multicommodity general equilibrium system. His formulation of the *tatonnement*, unlike modern formulations, was in terms of a sequential process where markets clear one at a time in a definite order. The first nonsequential approach to the multicommodity price-adjustment process was presented in Hicks (1939).

Let the general equilibrium of an n-good competitive economy be given by the n-equations:

$$z_i(p^*) = 0 \text{ for all } i = 1, 2, \ldots, n,$$

where $p^* = (p^*_1, p^*_2, \ldots, p^*_n)$ is the equilibrium price vector and $z_i(p^*)$ is the excess demand for i at these equilibrium prices. For such a system, Hicks defined "imperfect stability" of a particular market (say, i) as the case where $p_i > p^*_i$ implies $z_i(p) < 0$ and $p_i < p^*_i$ implies $z_i(p) > 0$, when all of the other (n - 1) markets are adjusted so as to stay in equilibrium. Hicks defined the "perfect stability" of market i as the case where these excess demand conditions hold whether the other markets are adjusted to remain in equilibrium or not. The n-good "system" is imperfectly stable (or perfectly) if all n markets are imperfectly stable (or perfectly).

In contrast to Hicks, the modern "dynamic" approach to the tâtonnement initiated with Samuelson (1941, 1942), where the price-adjustment mechanism was modeled as a system of first-order autonomous ordinary differential equations. For the n-good general equilibrium system, this "truly dynamic" adjustment is given by

$$dp_i/dt = \dot{p}_i = k_i z_i[p(t)] \text{ for all } i = 1, 2, \ldots, n,$$

where $p(t) = [p_1(t), p_2(t), \ldots, p_n(t)]$ is the prevailing price vector at time t, and $z_i[p(t)]$ is the excess demand for good i at price vector p(t), and $k_i > 0$ is the "speed of adjustment" for the i-th market. Simply put, prices increase for goods with positive excess demand and decrease for goods with negative excess demand. If p^* is an equilibrium price vector, then the local dynamic stability of p^* means that there exists a neighborhood around p^* such that for any initial price vector in this neighborhood the adjustment process [p(*t*)] converges to p^* as t→∞.

The relation between these two stability conditions—Hicksian and dynamic—has been particularly important to the development of general equilibrium theory. Based on the "correspondence principle" of Samuelson (1942), comparative statics results should follow from the assumption that the system is dynamically stable; and similarly, knowledge of the comparative statics properties of a system should be sufficient to determine its dynamic stability.

In his original presentation, Samuelson (1941) provided a crucial counterexample demonstrating that Hicksian perfect (and therefore imperfect) stability was not necessary for local dynamic stability of a general equilibrium system. This counterexample implicitly falsified the first part of his correspondence principle: that dynamic stability necessarily implies determinant comparative statics. Why? The reason is simply that the mathematical properties required to obtain comparative statics results crucially depend on the Hicksian conditions.[4]

The second part of the correspondence principle—that comparative statics information should provide information regarding dynamic stability—was equally damaged a few years later in Samuelson (1944), where a second crucial counterexample demonstrated that Hicksian stability was also not sufficient for dynamic stability. Together these two counterexamples proved unequivocally that Hicksian stability is neither necessary nor sufficient for the dynamic stability of a general equilibrium system.

Because of the intimate mathematical relationship between the Hicks conditions and the qualitative information obtained from comparative static analysis, these two counterexamples essentially destroyed the hopes of the correspondence principle holding for Walrasian models in general.[5] This negative result, while certainly not a falsification in any sense acceptable within the philosophy of science, has initiated a major redirection in the Walrasian research program. Modern commentators such as Arrow and Hahn now simply assert that "the correspondence principle 'isn't'" (1971, p. 321).

Even though the link between Hicksian and dynamic stability was broken in the most general case, it was long suspected (though unproven) that, if individual preference had the "normal" characteristics (indifference curves convex to the origin), then at least for a pure exchange economy the equilibrium price vector would be locally stable. Aberrant preferences were believed to be the only cause of instability; that is, "normally" a slightly disturbed price system would quickly return to an equilibrium position. A variety of results were obtained for special cases,[6] but no general result appeared to verify the suspicion that stability was the normal case.

Crucial counterexamples in Scarf (1960) and Gale (1965) proved these expectations wrong. It was shown that, even for "reasonable" economies with unique equilibria, there may be dynamic instability. The unique p* may be a "source" for the dynamic system (rather than a "sink"), sending prices speeding away from p* for even the slightest disturbance from equilibrium.

Gale's counterexample demonstrated that if any good was "Giffen"— that is, if $\partial z_i / \partial pi / > 0$ for any i—the adjustment process $\dot{p} = kz(p)$ could be "destabilized" for some choice of adjustment speeds $k = (k_1, k_2, \ldots k_n)$. The example by Scarf, on the other hand, did not require any good to be Giffen; instead, he assumed that all goods were normal and generated from quasi-concave utility functions. The goods in Scarf's example were extreme complements, so much so that all "substitution effects" were absent. This complementarity generated instability for Scarf much as the presence of a Giffen good had generated instability in Gale's example.

These two sets of crucial counterexamples initiated a major "problem shift" in the general equilibrium research program. As stated above, the first set of counterexamples by Samuelson severed the correspondence principle between stability and comparative statics for all but certain special classes of models. Thus stability alone is neither necessary nor sufficient for comparative statics results. The second set of crucial counterexamples restricted stability itself to special cases. It is only under very restrictive assumptions, such as gross substitutes, that the stability of general equilibrium can be guaranteed. During the years that followed the publication of these papers, research moved away from the study of the implications of stability to the study of the few particular cases in which it obtains. Another effect of these counterexamples was to accentuate the similarities between Walrasian general equilibrium models and other areas (or demi-cores) in economic theory—particularly growth theory, where equilibrium positions are notoriously unstable. Finally, the search for stability moved to alternative formulations of the price mechanism, such as "nontatonnement" or "non-Walrasian" equilibria.[7] These mechanisms violate the Walrasian no-trading-at-equilibrium assumption and have been applied in the area of microfoundations of macroeconomics, since such price mechanisms bear a family resemblance to certain disequilibrium macroeconomic theories.[8]

METHODOLOGICAL IMPLICATIONS

Some readers may feel that, while the above reconstructions are of some historical interest, they are without methodological importance. This potential criticism deserves a response since there are in fact a number of reasonable arguments that point to such a conclusion.

The first and probably most common of these critical arguments is that traditional methodological virtues—particularly the ability to survive severe empirical test—play a much more important role in a theory's development than these nonempirical counterexamples. In the presence of such traditional virtues, these counterexamples become merely ways of checking the accuracy of our mathematical instruments, and they are therefore without any deep methodological significance.

The problem with this argument is that, for many areas of analytical economics—general equilibrium being the paradigm case—such traditional methodological virtue does not exist. An exhaustive list of reasons for including general equilibrium among the more viable research programs in economics would not include its ability to survive the type of severe empirical test a falsificationist philosopher of science would find acceptable.[9] It was stated above that crucial experiments have been few in the history of economics; for mathematically sophisticated areas such as general equilibrium theory, "few" must be changed to "nonexistent."

No doubt some will protest this empirical indictment of one of the most prestigious areas of the discipline, but nonetheless the indictment seems to be the overwhelming consensus among recent methodological commentators.[10] Such an indictment *need not imply* that these theories are unimportant or uninteresting or unworthy of the intellectual resources committed to them.[11] In fact, once the old methodological myths are abandoned, the door is open for detailed investigations into the specific characteristics of such theories and the phenomena to which they were designed to refer.[12]

A second critical argument against the methodological importance of nonempirical counterexamples is that they do not entail a choice between two different theories as does a true crucial experiment. This criticism clearly cannot be applied to the above case studies, since in both of these cases there actually was a choice between two different theories. In the first case, the choice was between a theory where comparative statics predictions could be made whenever stability was present, and one where comparative statics predictions could be made only with the addition of ad hoc initial conditions. In the second case, the choice was between a theory where stability was always present, and one in which it was only present in certain special cases.

The third and probably most telling argument against the methodological importance of these two crucial counterexamples is the claim that such counterexamples are merely mathematics. It can be argued that these counterexamples merely show that certain conclusions do not follow deductively from certain antecedents, that is, that certain theorems are not really theorems at all. Such deductive failures among the empirically uninterpreted terms of the theory have no bearing on the theory's cognitive worth or empirical standing.

Such an argument clearly raises important questions about the growth of "knowledge" in the formalized areas of economic theory. If there is no direct empirical testing and no "external" crucial experiments for such areas, and if theory change is initiated by nonempirical counterexamples that are merely mathematics, then what demarcates such theories themselves from mere mathematics? This question is probably *the* important question for the history and philosophy of analytical economics, since it is not only general equilibrium theory, but also growth theory, as well as most of the important theoretical achievements of the past one hundred years,[13] that stand challenged by such questions.

Certainly the two case studies discussed above cannot provide adequate answers to such difficult questions. Much more (and more detailed) historical work needs to be done before these complex issues can be sorted out. It may be that, for these areas of economic theory, progress occurs in a way more similar to growth in mathematics than in natural science. Or it may be that theory change in some areas occurs in ways that are entirely unique to abstract economic theory.

Whatever the outcome of further investigation, at least one thing seems to be certain at this point: the correct approach to these questions is historical. To understand the nature of economic science it is best to start with the actual history of economics.[14] While this prescription may seem obvious, it is in fact a break with tradition. The traditional approach to understanding the growth of economic knowledge or economic theory change was to borrow methodological prescriptions from philosophers of natural science and then appraise (or usually condemn) economics on the basis of these borrowed standards. This stands in stark contrast to the historical approach taken above, where answers are to be found in detailed and careful examinations of the actual history of economics. While such an approach is not without its difficulties,[15] it seems to be the only reasonable approach to the complex questions of theory change and the growth of economic knowledge. Such an approach, of course, places a heavy responsibility on historians of economic theory—a responsibility that did not exist with earlier approaches to these methodological questions.

NOTES

1. Some philosophers of science argue that crucial experiments do not really exist in any science, natural or social. For instance, Imre Lakatos ar-

gues, " 'Crucial experiments' in the falsificationist sense do not exist: at best they are honorific titles conferred on certain anomalies long after the event" (1974, p. 320). While I am somewhat sympathetic to this view, a much weaker empirical claim is being made in the text. First, the assertion just made in the text concerns only economics; and second, it only claims that there have been "few, if any, such crucial experiments." Regarding this weaker claim, there seems to be a fairly strong consensus among economic methodologists (see Blaug, 1980a, for instance). A detailed discussion of the author's views on various aspects of the relation between economic methodology and recent philosophy of natural science is presented in Hands (1979) and (1984a). Since these intricacies are not at issue here, the current work is presented "as if" there existed a generally accepted (falsificationist) interpretation of the role of crucial experiments in natural science.

2. Such changes are "in some respects analogous to the physical scientist's crucial experiment" (Goodwin, 1980, p. 614).

3. Crucial counterexamples are certainly not restricted to "orthodox" economics. Ian Steedman (1975) provided such a counterexample to Michio Morishima's "fundamental Marxian theorem" that a positive rate of exploitation is both necessary and sufficient for a positive rate of profit under the capitalist mode of production (Morishima, 1973, pp. 53-54). Morishima countered the counterexample and surveyed the history of the theorem in Morishima and Catephores (1978, pp. 29-58).

4. Formally these stability conditions amount to restrictions on the excess-demand Jacobian matrix evaluated at the equilibrium prices. It is not necessary to discuss the mathematical details here; the argument is clearly presented in a number of places—Quirk and Saposnik (1968, pp. 195-216), for instance.

5. The word *general* is necessary here, since the link does exist for certain special cases such as "gross substitutes." These cases are discussed in Quirk and Saposnik (1968).

6. Metzler (1945), Morishima (1952), Negishi (1958), and Hahn (1961).

7. Hahn and Negishi (1962) and Fisher (1976), for instance.

8. Clower (1965) and Barro and Grossman (1976), for instance.

9. The question of whether any other science meets such a criterion is simply not at issue here (see note 1 above).

10. Blaug (1980a), Clower (1975), Coddington (1975), Hausman (1981a), and Rosenberg (1980), for instance.

11. Although some do argue this way—Blaug (1980a), for instance—this is not a view that I share.

12. For instance, see Coddington (1975), Hausman (1981a), Weintraub (1983), and various papers in Stegmuller, Balzer, and Spohn (1982).

13. Including recent formalizations of Marxian economics (see note 3 above).

14. See Hausman (1980).

15. Some of these difficulties are discussed in Hands (1984a).

3

Blaug's Economic Methodology

This essay (Hands, 1984a) on Mark Blaug's Methodology of Economics *(1980a) originally appeared in* Philosophy of the Social Sciences. *During the methodological revival of the 1980s, Blaug's book, along with Bruce Caldwell's* Beyond Positivism *(1982), became a standard reference in the methodological literature. Both of these books exhibited the type of philosophical understanding that had been present in the methodological classics of the late nineteenth and early twentieth centuries, but seemed to disappear from the frequently cited methodological writings of the post-World War II period. Despite my appreciation of Blaug's general approach, I had (or have) a number of criticisms regarding the particulars of his methodological position; these criticisms not only appear in this chapter but in a number of the following chapters as well.*

Mark Blaug is one of the few economists whose perspective on economic methodology has been directly influenced by recent developments in the philosophy of natural science.[1] For example, Blaug was the first economist to introduce the work of the philosopher of science Imre Lakatos to the American economic profession. He is also unique for including a detailed discussion of current philosophy of science in his popular history of economic thought textbook.[2]

The object of this essay is to examine Blaug's most recent contributions to the field of economic methodology.[3] Blaug (1980a) is a general survey of the field of economic methodology, while Blaug (1980b) is a Lakatosian appraisal of classical Marxian economics. Because of its more general interest, the discussion will focus mainly on Blaug (1980a), although the two could easily be regarded as one comprehensive work. Since Blaug (1980a) ends with an appraisal of neoclassical economics and Blaug (1980b) is relatively brief, the latter could easily have been appended to the former, thus creating one complete book on economic methodology appraising both the neoclassical and the Marxian research programs.

Blaug (1980a) is divided into three relatively distinct parts. Part one (chs. 1 and 2) is a survey of the philosophy of science, with a particular emphasis on the developments of the past decade or so. Part two (chs. 3-5) is a descriptive history of methodological discussion in economics, and part three (chs. 6-15) is an evaluation of modern economics in the light of the preceding methodological discussion. The chapters of the last part are relatively short since each focuses on one particular topic in economic theory and/or its application.

In what follows I will discuss each of these three parts in sequence. For the first two sections the critical comments are of a relatively minor sort, concerned mostly with emphasis or priorities rather than substantive issues. The final section on the other hand, offers an alternative to Blaug's interpretation of the relation between economic methodology and modern economic theory. Prior to the discussion of Blaug's first part I will present a survey of recent developments in philosophy of science.

PHILOSOPHY OF SCIENCE

Any responsible discussion of philosophy of science must recognize that the discipline has undergone a major upheaval during the past twenty years. The so-called received view of the preceding epoch is dead.[4] The chorus of dissident voices that arose against it emanated from every conceivable corner of philosophic scholarship. These disparate voices have yet to harmonize into what should be called a "new view," but the dominant characteristics of such a consensus are now quite clear.

First and foremost, the new view will be a philosophy of *science*. That is, it will be a philosophy based on the actual historical practice of science, rather than, as previously, an analysis of an idealized structure bearing only a fortuitous resemblance to actual scientific practice.[5] This historical turn is by no means favored by every philosopher of science,[6] and there is still disagreement even regarding the appellative conventions;[7] but there is little doubt that any philosophy of science gaining influence in the immediate future will do so by proving itself in the crucible of history. This view is summarized nicely in a survey by Frederick Suppe:

Today virtually every significant part of the Positivistic viewpoint has been found wanting and has been rejected by philosophy of science. The dominant thrust of contemporary work in the philosophy of science is the development of new views of science which proceed from, and tend to depend heavily on, close examination of actual practice and products (e.g., theories, explanations) of science. (Suppe, 1979, p. 317)

This historical turn is not without its problems. Of the two most important, the first could be called the "circularity problem"; it is a problem that arises in the following way. If we are going to derive our philosophy of science empirically, we must examine the "facts of science." But what are these "facts"? How do we know we are observing "science" and not something else? We must have a demarcation criterion—something that allows us to discriminate scientific activities from nonscientific activities. The problem is, the demarcation criterion actually determines the data to be examined, and in turn determines the view of science extrapolated from that data. Thus the philosophy of science, obtained "empirically," will be merely a reflection of the initial demarcation criterion employed. This circularity may or may not be an insurmountable problem for the empirical philosophy of science.[8]

The second basic problem is the problem of "irrationality." To explicate this particular usage of the term *irrational* it is helpful to make an aside into the empirical history of science. Traditional philosophy of science established norms—rules that decreed how science "ought" to proceed. Depending on exactly which traditional school is adhered to, following the rules will produce something "meaningful," or "truth," or something with "verisimilitude," but something called "science" and something substantially different from nonscience in any case.

There is nothing a priori to tell us whether an examination of the actual history of science will uncover activities that conform to these norms or not. A priori, it could be that actual scientists are much more conservative than the norms allowed—seldom if ever venturing to the boundaries of permissiveness. On the other hand, (a priori) it could be that actual scientists behaved in just the opposite way, violating the rules with impunity at every turn. Which of these two views is supported by the history of science? Regrettably, it seems that recent work in the history of science gives unequivocal support to the latter view. Particularly the influential work of Paul Feyerabend and Thomas Kuhn has demonstrated that the history of "great science" is a history of breaking the rules.[9]

In this way a new constraint enters the problem of choosing between traditional philosophy of science and a more historical alternative. It seems history-based philosophy of science will always be more liberal, more permissive, more likely to accord more types of activity the status of science. The problem of irrationality develops because the norms of traditional philosophy of science are norms of rationality; they provide an objective reason for recommending a particular type of scientific behavior. The result of recent empirical work, however, is that no general norms emerge from the historical studies of science. Theories are accepted or rejected not on the basis of the objective criteria or any traditional philosophy of science, but on the basis of sociological or psychological habits and conventions. Scientists are either not following any rules at all, or the rules they are following seem to be external to the activity of "science." Science thus loses its epistemic uniqueness; it is no more rational than any other type of sociological/psychological-group-oriented behavior. It is in this sense "irrational."

One of the major influences in philosophy of science during the past ten years has been the work of the late Imre Lakatos. Lakatos addressed himself directly to the question of irrationality. He tried to formulate a methodology that would be compatible with the actual history of science *as well as* render that history rational. For Lakatos, failure to establish such a methodology would mean that "we have to declare that the most important, if not all, theories ever proposed in the history of science are metaphysical, that most, if not all, of the accepted progress is pseudo-progress, that most, if not all, of the work done is irrational" (1970, p. 103). Lakatos wanted to preserve the critical bite of tradition,[10] while testing his methodological prescriptions against history.[11] He sought the best of both worlds, a way "for the philosopher of science to learn from the historian of science and vice versa" (1971a, p. 111).

How successful was Lakatos in achieving his goal? It is still too early for a definitive answer to this question, but at this point things do not look particularly promising for the Lakatosian position. The old adage that "the person who walks in the middle of the street gets hit by cars going both ways" may apply. Representatives of the recent historical turn regard Lakatos as neglecting or, worse, forgetting history.[12] More traditional philosophers of science argue that he has conceded too much to history and so the "normative content evaporates."[13] In addition, there is the "Popper versus Lakatos" issue. Since Lakatos's work emanates from the Popperian tradition, there is a question of his fidelity. Has Lakatos fortified the Popperian position by solving some of its difficulties, or merely presented a spurious caricature of the Popperian original?[14]

These questions surrounding Lakatos's work are by no means settled. Probably the best that can presently be said regarding his position is that it was "a good stab,"[15] which tried "to do too much."[16] The important point is that he addressed the right *question*. He was undaunted in his desire to accommodate the current historical turn without sacrificing traditional virtue; he sought "a defensible intermediate position."[17]

After this rather long introduction, it is time to examine Blaug's interpretation of the history of the philosophy of science. Is the above story the one Blaug tells? The answer is basically yes.

Blaug surveys the "received view" (pp. 1-9),[18] with particular attention to its "normative" characteristics. He argues that such a normative view is "somewhat out of date" (p. 1), and that the philosophy of the received view "excludes much of what at least some people have regarded as science" (p. 9). This failure to "tell it like it is" characterizes what "its critics find so objectionable" (p. 9). He explains that this "tension" between description and prescription "has dogged the received view of scientific theories for over a generation" (p. 33) and has "been a leading factor in the virtual overthrow of the received view" (p. 10).

Popper is presented as "one of the mainsprings of the opposition to the received view" (p. 10) and as a "watershed between old and new views" (p. 2). The fundamentals of Karl Popper's position are presented, as well as the most frequent criticisms, such as the Duhem-Quine nonfalsifiability thesis (pp. 17-19), the theory-ladenness of observations argument (pp. 14 and 41-42), and the closely related "conventionalist" (p. 20) element in Popper's philosophy of science so frequently discussed by Lakatos.[19] Blaug is correct in emphasizing that "Popper's methodology is plainly normative, prescribing sound practice in science, possibly but not necessarily in the light of the best science of the past" (p. 29).

Lakatos is presented as a "somewhat mischievous" (p. 26) interpreter of Popper. Lakatos's desire for an intermediate position on the historical versus normative issue is clearly stated. Lakatos presents "a compromise between the ahistorical, if not antihistorical, aggressive methodology of Popper and the relativistic, defensive methodology of Kuhn" (p. 34). Blaug closes his section on Lakatos with a quite realistic appraisal of the Lakatosian position. He states that Lakatos's effort "to retain a critical methodology of science that is frankly normative, but which nevertheless is capable of serving as the basis of a research program in the history of science, must be judged either as a severely qualified success or else a failure, albeit a magnificent failure" (p. 40).

Blaug gives an equally adequate rendering of the recent historicist turn. He not only discusses Kuhn (pp. 29-33), Feyerabend (pp. 40-44), and "the new view on scientific theories" (p. 41), but he also makes the same type of cautiously sympathetic statements that appear so often in recent philosophy of science literature. "To preach the virtues of *the* scientific method, while utterly ignoring the question of whether scientists now or in the past have actually practiced that method is surely arbitrary" (pp. 33-34).

From this it appears that Blaug's overall construction of the recent history of philosophy of science is an accurate portrayal of the philosophical literature. In closing this discussion of his part one, I would like to make one additional comment regarding the evolution (at least in print) of Blaug's ideas on these topics. It appears that as the literature of Lakatos—much of it critical—has accumulated, Blaug's enthusiasm for the Lakatosian position has waned slightly. For instance, the reader is now told, "It is fair to say that Lakatos in the final analysis has the same difficulty that Popper experienced in steering a middle course between prescriptive arrogance and descriptive humility" (p. 39). In his earlier discussion of these topics (Blaug, 1975) the emphasis was more positive, the reader was told that Lakatos's work amounted to "virtually resolving a long-standing puzzle about the relationship between positive history of science and normative methodology for scientists."[20] This support of Lakatos prompted one reviewer of Blaug (1975) to label it a Lakatosian "diatribe."[21] Such criticism is certainly an exaggeration. Blaug (1975) was not a Lakatosian diatribe, but it did appear marginally more sympathetic to Lakatos than Blaug's later work.

ECONOMIC METHODOLOGY

In his second part (pp. 55-156), Blaug presents a descriptive history of economic methodology. Such a history has not been available prior to Blaug's contribution. Most discussions of economic method contain a brief presentation of conflicting and/or preceding views, but no comprehensive historical survey of the topic has been available. It is important to emphasize that economic methodology refers to writings about the methodology of economics—that is, philosophy of economic science—and not necessarily the method employed by practicing economists.

Blaug provides a very comprehensive history of the subject. He details the methodological prescriptions of the English classical school—an arduous task, since it is an area particularly sparse in secondary sources (until now). He discusses in detail each of the methodological classics of the twentieth century: the a priorism of Lionel Robbins, the dogmatic falsificationism of Terence Hutchison, as well as the more recent methodological "contributions" of Milton Friedman and Paul Samuelson. All in all, Blaug's descriptions are quite good and leave little room for criticism.

One nontrivial exception is his two-page treatment of Austrian methodology (pp. 91-92). Blaug is probably correct in being critical of Austrian methodology,[22] but there are a number of reasons why the position deserves more attention than it is given. First, the most extreme Austrian view is a methodology built on "synthetic-a priori-truths." While this is not currently a mainstream philosophical position, it is a view with a long philosophic ancestry and it certainly deserves more attention than Blaug has given it. Second, the Austrian view is the only major position in economic methodology that is radically against "methodological monism" (p. 64), arguing instead that the methodology of economics is fundamentally different from the methodology of natural science (i.e., different in *type*, not only *degree*). Third, there is a potential connection between the Austrian view and Karl Popper's view of economics. This point really warrants a separate and complete study, but there is reason to believe that F. A. Hayek provides a missing link between the Austrian position and Popper's views on economic methodology. In any case, the question of Austrian methodology deserves more coverage if Blaug's description of economic methodology is to be complete.

The final chapter of this part on the history of economic methodology contains a discussion of the distinction between positive and normative economics. Since this is an area in which the economic conventional wisdom is badly in need of modernization, Blaug's arguments are extremely important and will be discussed in much more detail than the other topics in this section.

Almost every textbook on economic theory—regardless of the level of analytical rigor or intended audience—begins with the demarcation between "positive" and "normative" economics. Positive economics tells us what *is*, and normative economics tells us what *ought* to be. After presenting these definitions, most authors then assert that their own text contains exclusively positive economics. On first exposure to such statements, the philosophically inclined reader is left with the feeling that the epistemic warrant for having made them is something that will be provided later—being postponed, possibly for heuristic reasons, until after the rudiments of the theory itself have been presented. Such warrant is never provided, however. Nothing more is ever said on the topic; the distinction and the claim are merely repeated and reprinted ad infinitum (and ad nauseam).

If economics tells us what *is*, it must reach unerringly into the reality behind the appearances; it must transcend the illusion of the Platonic shadows and expose the true causes of the economic phenomenon we observe. While such unequivocally true scientific theories might be a desirable goal, most philosophies of science—even those of the realist persuasion—do not argue for such a rigid characterization of scientific theories. Most logical positivists only claimed positive knowledge of subjective sense experiences. Popper has always argued for "fallibilism" against the position that science, no matter how well practiced, tells us what "is" in some absolute essentialist sense. We can never be entirely certain that any scientific theory, and thus certainly not an economic theory, tells us what is.

Regarding what *ought to be*, Blaug argues (consistent with contemporary philosophy of science) that observations are "theory laden." Given theory-ladenness, the acceptance or rejection of a particular scientific theory will always involve a judgement—albeit a methodological judgement, not a moral judgement. As Lakatos argued so often, "It is not that we propose a theory and Nature may shout NO; rather we propose a maze of theories and Nature may shout INCONSISTENT" (1970, p. 130). Given an observed inconsistency between "the data" and "the theory," the choice of action necessarily entails a judgement. Do we take the observation theories as unproblematic background knowledge and reject the theory (as would Popper)? Do we hold the theory as beyond doubt and assume that we the scientists have made a mistake (as in Kuhnian normal science)? Such a decision rests, Blaug explains, "on the willingness to abide by certain rules of the game, that is, on judgements that we players have collectively adopted" (p. 132). These decisions are made by "convention," and the process employed in reaching agreement on these methodological conventions "is not a very different cognitive process from the acceptance or rejection of ought statements" (p. 131).

Blaug's discussion is a good beginning for a badly needed critique of the positive/normative dichotomy as commonly used in economics.[23] Regrettably though, Blaug ends up diluting his argument by claiming that the distinction "should be clearly maintained as far as it can be" (p. 140). This philosophic atavism is preserved ostensibly because to do otherwise would "take us straight into the camp of certain radical critics" (p. 134).

This argument seems to be below the standards of the rest of Blaug's work. Economics either tells us what is, or it does not; who benefits from a particular answer should be irrelevant to the argument. Besides, no "radical" alternative has any better epistemic standing in this respect. If Blaug feels the economic conventional wisdom on this issue is acceptable, why was the (excellent) critical argument made at all? Let us hope that it is not because he, as well as his philosophically attuned readers, are capable of responsibly handling such knowledge about the positive/normative distinction, while the masses and the less philosophically mature economists are not.

Regardless of Blaug's efforts to backpedal from his own results, or his reasons for doing so, his original point is well taken. The positive/normative distinction as used in standard economics texts is philosophically naive at the very best. The topic certainly deserves more attention than it can be given here, but hopefully Blaug's argument (and similar positions recently taken by other economists) will initiate a long overdue reexamination of the positive/normative distinction in economics.

THE APPRAISAL OF MODERN ECONOMICS

The third task that Blaug sets himself is much more evaluative and much less descriptive than the tasks of the first two parts. Armed with the recent work in philosophy of science, as well as the history of economic methodology, Blaug makes a methodological evaluation of the conventional wisdom of economic science. What do economists do? How does what they do stand up to the probing light of methodological appraisal? Blaug's answer to these questions, in short, is that economics does not stand up very well at all.

The heart of neoclassical economics—static demand theory—is not falsifiable (pp. 162, 165-67, 169). The theory of competitive firm behavior (long the brunt of methodological criticism) is not tested in the static case (p. 186), and not even testable "in principle" in the dynamic (p. 184). Growth theory is equally indictable: its beautiful dynamics are merely "brain twisters" (p. 254), "devoted to logical puzzles" (p. 255), and "not as yet capable of casting any light on actual economies growing over time" (p. 254). The great Keynesian-monetarist macropolicy debate is similarly unspared: it "must rank as one of the most frustrating and irritating controversies in the entire history of economic thought, frequently resembling medieval disputations at their worst" (p. 221).[24]

Walrasian general equilibrium (GE) theory (long the apogee of disciplinary prestige) receives the most scathing attack of all. It "makes no predictions" (p. 188), "has no empirical content" (p. 189), "does not claim to describe the real world in any sense whatsoever" (p. 190); it lacks "any bridge by which to cross over from the world of theory to the world of facts" (p. 191) and, finally, "is at best a species of 'solving the puzzles that we have ourselves created,' and time spent in mastering it is time taken away from learning the empirical methods of economics" (p. 256). This criticism is particularly telling to the whole neoclassical program since, as Blaug himself stated in another context, "nearly all economics nowadays *is* Walrasian economics" (1978, p. 617).

Blaug's methodological wrath is not restricted to only orthodox economics. Blaug (1980b) makes similar and even stronger indictments of classical Marxism. An absence of "empirical content" is also credited to the economics of the new left (p. 258), the Cambridge (U.K.) program (p. 257), American institutionalism (p. 259),[25] as well as the neo-Austrian program (p. 259). It would seem that only econometrics is spared; but even it, in its present form, "quickly degenerates into a sort of mindless instrumentalism" (p. 257).[26]

Blaug's overall appraisal of modern economics is best given by his statement that "modern economists frequently preach falsificationism, . . . but they rarely practice it" (p. 129).[27] He states in summary that his "own contention . . . is that the central weakness of modern economics is, indeed, the reluctance to produce the theories that yield unambiguous refutable implications, followed by a general unwillingness to confront those implications with the facts" (p. 254).

Blaug's criticism is by no means unique; most methodological commentators have come to similar conclusions, especially with respect to general equilibrium theory.[28] Some of these commentators are criticizing the theory in an attempt to destroy it and replace it with an entirely different program, while others—and Blaug is in this second camp—merely seek deep and fundamental methodological reform.[29] Given the massive amount of literature that comes to similar conclusions (although most of this literature is simply ignored by the economics profession), Blaug's empirical claim regarding the lack of testing and testability in modern economics should be accepted. But granting Blaug's empirical claim is by no means sufficient for the acceptance of his negative appraisal of modern economics.

The fundamental problem with Blaug's criticism of economic practice is that it either completely neglects, or at least is inconsistent with, everything he stated in his survey of philosophy of science. His principal criticism is that there is not enough "falsification" or even "falsifiability" in modern economics. In light of the previously discussed work of Kuhn, Feyerabend, Lakatos, and others, one is inclined to retort— *so what?* The unambiguous conclusion of recent philosophy of science, which Blaug so accurately surveys in part one, is that *no science* does that which Blaug criticizes economics for not doing. If the actual history of any science were a history of falsifications or even attempted falsifications, there would have been no Lakatosian program. The "new view" is a direct response to the general lack of falsificationist practice in the history of natural science. Chastising economists for failure to follow a falsificationist methodology is not methodological monism; instead it amounts to foisting a methodology on economics that even the best gambits of physical science have failed to live up to.

It is possible merely to deny the literature of the historicist turn, as many Popperians have done, and assert that methodological norms cannot be tested against the actual history of science.[30] While this is becoming a minority view in philosophy of science, it is certainly a position Blaug could have taken in part one. The point is, this is not the position he did take, and therefore his criticism of economics is inconsistent at best.

In many respects, the inconsistency problem in Blaug's appraisal of economics is minor compared to another problem posed by his position. His appraisal forces economics into the same pitfall that Lakatos feared for natural science; that is, it renders the growth of economic knowledge "irrational." Since economics does not conform to strict falsificationist prescriptions, its actual history is left open to be explained by "class allegiances," "bourgeois ideology," "collectivist statism," or some other interpretation that attributes the choice between competing economic theories to something extrascientific and "external" to the scientific endeavor itself. As Lakatos argued for physical science, such charges can only be avoided if the methodology proposed is one that renders at least "most" of the history of economics "rational." Blaug's does not.

The question must be asked—why? Why does Blaug, who obviously—from the elegant presentation in the first part—has an excellent understanding of current problems in philosophy of science, want to foist the philosophic oldspeak of falsificationism on economics? Probably the motivation is fear, that is, the fear of abdicating the sanctity of scientific economics to those pamphleteers of extreme political persuasion who inhabit the underworld of economics. After all, if we lower our empirical standards, won't anything be able to pass as economic science?

The problem is analogous to an old problem in statistical theory. If one wishes to avoid a type-II error (accepting a false hypothesis) at any cost, the obvious solution is simply to accept nothing. The problem is that this leaves the probability of a type-I error (rejecting a true hypothesis) quite high. Strict falsificationist rules certainly keep the riffraff out of science, but they also keep most of science out of science. In the case of economics, Blaug's methodology leaves the greatest economic theorists with no more sanctity than the most trivial of pamphleteers, since both fail to meet his standards. This is no more than Lakatos's central problem of irrationality.

It seems that Blaug has learned his philosophy quite well but has fallen short in applying this learning on his home court. Blaug is not alone among economists in this respect. As Keynes warned so many years ago, in economics "the difficulty lies, not in the new ideas, but in escaping from the old ones, which ramify, for those brought up as most of us have been, into every corner of our minds" (1936, p. xxiii).

CONCLUSION

Even considering the criticisms of his appraisal of modern economics, Blaug has accomplished quite a lot. Blaug (1980a) is a book that hopefully will receive wide circulation within the economics profession. His part one is excellent; it represents one of the best surveys of its kind, and would qualify as a clear introduction to philosophy of science for a reader from any field. Part two is a long overdue descriptive history that has the potential to nurture research work on a variety of specific topics. Part three must certainly be given high marks for scholarship and detail. As was argued above, his appraisal is by no means the only one available, but it was conscientiously presented nonetheless. Overall this is good work, written with a degree of philosophic sophistication that is as refreshing as it is rare in the literature of economics.

NOTES

1. The list of all such economists, while short, does nevertheless include a few others (the late Alan Coddington, for one). It is important to note that the list does *not* include the names most economists think of when "economic methodology" is mentioned, e.g., Milton Friedman, Lionel Robbins, and Paul Samuelson, although Robbins (1979) may be an indication of change in this respect.

2. Blaug (1975), and "A Methodological Postscript," pp. 697-727 of Blaug (1978). Blaug has also made substantial contributions to economics itself, especially in the areas of human capital theory and economics of education.

3. Blaug (1980a) and (1980b).

4. The use of the *term received* view for the logical-positivist-based philosophy of science dominant in the interwar period was popularized by Frederick Suppe in Suppe (1977). His book contains an excellent summary of the received view as well as the arguments of its early critics.

5. Support for this claim can be found in almost any recent discussion of general philosophy of science. For particularly clear statements, see Laudan (1979), McMullin (1976, 1979), or Suppe (1979). The argument is made with specific reference to economics in Hausman (1980).

6. Patrick Suppes (1979, p. 16) refers to "the new imperialism of historical methods."

7. Larry Laudan (1979, p. 40) discusses "historical methodologies"; Ernan McMullin (1979, p. 55) calls it a "historicist turn" and later, p. 56, the "Kuhnian revolution"; for Frederick Suppe (1979, p. 319), they are "growth of knowledge theories"; and for Daniel Hausman (1980, p. 353), it is "empirical philosophy of science."

8. For instance, Merrill (1980) thinks it is, while Brown (1980) thinks it is not.

9. Feyerabend (1975a); Kuhn (1970a). For a popularized version, see Brush (1974).

10. "We *must* find a way to eliminate *some* theories. If we do not succeed, the growth of science will be nothing but growing chaos" (Lakatos, 1970, p. 108).

11. "A general definition of science thus must reconstruct the acknowledgedly best gambits as 'scientific'; if it fails to do so, it has to be rejected" (Lakatos, 1971a, p. 111).

12. Thomas Kuhn (1971, p. 143) refers to Lakatos as "not history at all but philosophy fabricating examples." Wolfgang Stegmuller (1978, p. 69) calls Lakatos's work a "philosophy of wishful thinking."

13. Kulka (1977, p. 341). Paul Feyerabend (1975a, 1976)—certainly no traditionalist—also argues that Lakatos is without normative bite, but for Feyerabend this is praise, not criticism. Kuhn and Feyerabend's criticisms are defended by Philip Quinn (1972).

14. Noretta Koertge (1978, p. 269) refers to Lakatos's work as an *inversion* of Popper's. Joseph Agassi, in a particularly invective discussion, states, "Had Lakatos learned Popper better, we would not have the Lakatosian school on our hands. Ignorance is bliss" (1979, p. 320). Popper himself is one of those who argue that Lakatos has taken liberties with his work: "I feel, unfortunately, obliged to warn the reader that Professor Lakatos has, nevertheless, misunderstood my theory of science: and that the series of long papers in which, in recent years, he has tried to act as a guide to my writings and the history of my ideas is, I am sorry to say, unreliable and misleading" (1974, p. 999).

15. Hacking (1979, p. 401).

16. Adler and Elgin (1980, p. 415).

17. McMullin (1979, p. 66).

18. Pages in parentheses refer to Blaug (1980a) unless specifically indicated otherwise.

19. And frequently ignored by economists; see Coddington (1972, pp. 10-11) and Hands (1979, pp. 295-98).

20. Blaug (1975, p. 150). Blaug (1975) is reprinted in Latsis (1976a), and page references are to that reprint.

21. Agassi (1979, p. 322).

22. Blaug states it is "wholly alien to the very spirit of science" (1980a, p. 93).

23. Blaug is certainly not the only economist to make this point regarding "methodological norms." See Coddington (1972), Gordon (1977), and Tarascio (1971), for example.

24. Similar remarks on each of these fields are contained in Blaug (1975) and "A Methodological Postscript," pp. 697-727 of Blaug (1978).

25. More detailed criticism of American institutionalism is given in Blaug (1978, pp. 710-13).

26. On the other hand, Blaug does claim that econometrics is "our only hope" (1980a, p. 261).

27. For similar "they don't practice what they preach" remarks, see Blaug (1980a, pp. xii-xiv, 113, 249, 259-60); also Blaug (1975, pp. 159 and 174), and Blaug (1978, pp. 697-98).

28. Robert Clower, for instance, states,

Strictly interpreted Neo-Walrasian theory is descriptive only of a fairy-tale world of notional economic activities that bears not the slightest resemblance to any economy of record, past, present or future. It is science fiction, pure and simple—clever and elegant science fiction, no doubt, but science fiction all the same. (Clower, 1975, p. 10)

For similar statements regarding the empirical standing of neoclassical economics generally, and GE particularly, see Coddington (1972 and 1975), Handler (1980a and 1980b), Hausman (1981a and 1981b), Hutchison (1976), Kaldor (1972), Latsis (1976b), and Rosenberg (1980).

29. For instance, Alexander Rosenberg states, "My purpose . . . is not destructive, for I consider the theory under discussion to be the most impressive edifice in social science yet erected" (1980, p. 79).

30. Popper himself certainly takes this position: "my theory is not empirical, but methodological or philosophical, and it need not therefore be falsifiable" (1974, p. 1010); similar remarks on p. 1036. This is the old positive versus normative problem now surfacing at the metamethodological level.

4

Second Thoughts on Lakatos

This paper (Hands, 1985a), unlike Chapter 1, was written after a number of Lakatosian case studies had appeared in the economics literature, and consequently its tone is substantially more critical of Lakatos than the earlier chapter. Since Weintraub (1985b) was not published at the time this essay was written, the criticisms of his interpretation of general equilibrium theory are based on earlier work—primarily Weintraub (1979). The discussion of Blaug's interpretation of the Keynesian revolution from this chapter continues into Chapter 5 as well.

TEN YEARS OF LAKATOSIAN METHODOLOGY

Over a decade has passed since the first applications of Imre Lakatos's methodology of scientific research programs (MSRP) to economics were presented at the Nafplion Conference in September 1974. During the intervening years, our ability to pass judgement on "Lakatosian economics" has improved in at least two important ways. First, during this time the Lakatosian program itself (now called the neo-Lakatosian program by some) has been greatly clarified and articulated. Because of Lakatos's untimely death, most of this articulation has been carried on by other members of the Lakatosian "school"—particularly John Worrall and Elie Zahar.[1] Although there is still not a consensus within the philosophy of science regarding the success of the Lakatosian program, there is now at least a fairly good consensus regarding its content. Second, our ability to evaluate Lakatosian economics has improved

37

because the economics profession has taken up the call to provide a number of Lakatosian rational reconstructions in the history of economic thought. In addition to the original Nafplion papers on Keynesian economics (Blaug, 1976a; Leijonhufvud, 1976), the theory of the firm (Latsis, 1976b), and the Leontief paradox (de Marchi, 1976), we now have applications to Marxian economics (Blaug, 1980b), human capital theory (Blaug, 1976b), monetarism (Cross, 1982), Austrian economics (Rizzo, 1982), financial economics (Schmidt, 1982), and general equilibrium theory (Weintraub, 1979).[2] While still far from professional household words, Lakatos's "pornographic metaphors" (Hacking, 1979, p. 398) of "hard core" and "protective belt" occur with ever-increasing frequency in the current economic literature.

In light of these two factors—the clarification of the MSRP itself, and the increased availability of suitable case studies—it is now time for a reevaluation of the applicability of the MSRP to economic theory. The discussion that follows is just such a reevaluation. The overall conclusion of the analysis, unlike most previous analyses of this question, is primarily negative. It will be demonstrated that the MSRP's strictly empirical criteria of "progress" makes a Lakatosian rational reconstruction of the most successful episodes in the history of economic thought virtually impossible. While this conclusion obviously has negative implications for future research in Lakatosian economics, the implications are not entirely negative. It will be argued that the fundamental problem Lakatos addressed in the MSRP is precisely the problem that still faces economic methodologists. In fact it could be argued that Lakatos's basic problem is not only present, but even more pronounced in economics than in natural science. In the final section it will be suggested that economics should not start with Lakatos's methodology itself—that is, not with his solution—but with his question, with the problem that originally motivated his methodology.

WHY THE ATTRACTION?

Why did methodologically inclined economists become interested in Lakatos's work in the first place? What is it about the Lakatosian program that has made it so attractive to economists?[3]

The first and probably foremost attraction is that the MSRP as a philosophy of science mitigates the importance of falsification or refutation in demarcating science from other types of activity. For Lakatos, "since all theories are born refuted, bare 'refutations' can play no dramatic role in science" (1968, p. 163). "Scientists . . . are not irrational when they tend to ignore counter examples . . . and follow the sequence of problems as prescribed by the positive heuristic of their program, and elaborate—and apply—their theories regardless" (Lakatos, 1970, p. 176). Such empirical permissiveness is particularly endearing to a discipline long characterized by "innocuous falsificationism" (Blaug, 1976a, p. 160). Regardless of how much the economics profession preaches falsificationism, it is the universal consensus of recent methodological commentators that they almost never practice it.[4] A methodology that allows access to the kingdom of science without repentance for a lifetime of nonfalsificationist practice is simply too alluring for many economists to resist.

A second important attraction for the MSRP is the emphasis on "internal" explanations of theory change. Lakatos's methodology is designed to attribute "problem shifts" within a research program as much as possible to the internal dynamic of the research program itself, rather than leaving theory change to be ascribed to "external" or extrascientific causes. Providing such internal explanations of theory change in science is one of Lakatos's basic motivations; his principal criticism of Thomas Kuhn's approach is that Kuhn makes theory change a matter of "mob psychology" (Lakatos, 1970, p. 178) and therefore external or exogenous to any model of the rational growth of scientific knowledge. Economics has long been accused of being "externally" driven. For many, the history of economic theory is a history principally, if not exclusively, driven by such things as the "bourgeois class interest" or, from the other side of the political fence, an "anticapitalistic mentality." If the MSRP can be used to immunize economics against such age-old criticisms, particularly without forcing the profession to change its nonfalsificationist ways, Lakatos may be the methodological redeemer the profession has long awaited.[5]

A third, and perhaps less important, reason for the attraction is that the MSRP allows for so-called Kuhnian loss during a scientific revolution. Frequently during a scientific revolution the succeeding theory simply fails to address a particular class of phenomena that could be explained by the preceding theory. Cases of such Kuhnian loss abound in economics. For instance, the movement from classical to neoclassical economists involved a Kuhnian loss with respect to population theory. Within classical economics, human population was an endogenous variable, something to be explained by the theory. Neoclassical economics did not offer a different theory of endogenous population; it simply offered no theory of population at all. Population simply fell outside the domain of things that the theory explained. Most philosophies of science have great difficulty accounting for how it is possible for a new theory to constitute *progress* over an older theory in the presence of such losses. This difficulty does not arise with the MSRP. On the basis of Lakatos's criteria of "progress," it is entirely possible for a successor to constitute a "progressive problem-shift" over its predecessor even in the presence of such losses between the two theories.[6]

Finally, the MSRP recognized explicitly that scientific theories are dynamic and constantly evolving entities. It is a fundamental tenet of the Lakatosian view that the principal products of the scientific enterprise are scientific research *programs*, not merely specific scientific theories. A moment's thought about international trade theory, or the quantity theory of money, or general equilibrium theory will reveal how consistent this view is with the history of economic thought. As John Worrall put it, "It is an historical fact that some important theoretical innovations in science were quickly succeeded by the articulation of a series of scientific theories, related in certain ways to, but not implied by, the first theory" (1978, p. 58).

APPRAISING METHODOLOGIES

The central purpose of Lakatos (1971a) was to provide a "meta-methodology," that is, a general method for appraising methodologies. Lakatos states that "a general definition of science thus must reconstruct the acknowledgedly best gambits as 'scientific': if it fails to do so, it has to be rejected" (Lakatos, 1971a, p. 111).[7] Thus, a methodology—a view about the nature of scientific rationality—is preferred if it can "internalize" or "rationalize" a larger portion of the actual history of that which is generally acknowledged to be the "best" science. How are these "best gambits" determined? They are determined by the scientific elite itself: "if a demarcation criterion is inconsistent with the 'basic' appraisals of the scientific elite, it should be rejected" (Lakatos, 1971a, p. 111).[8] Lakatos's position is based on the simple argument that, while there is no agreement on what constitutes science in general, there certainly is agreement on the scientific standing of certain specific cases.

If we are to follow this metamethodology in appraising the MSRP as a methodology of *economic* science, it appears that we should proceed by examining the best gambits of economics through Lakatosian spectacles. If the acknowledged best economic theories appear rational in the light of MSRP—that is, if the profession's acceptance of these theories can be explained internally by the standards set forth in the MSRP—then Lakatos's methodology should be given positive marks with respect to economic science. Thus, at least in a certain restricted sense, case studies in the history of economic thought provide a test of the adequacy of the MSRP for economics. The word *restricted* is necessary here because we want to avoid falsificationism creeping in at the meta level. Just as Lakatos would argue that an anomalous observation does not impel us to abandon a scientific theory, an anomalous observation from the history of the best science does not force us to throw out a particular view of scientific rationality. On the other hand, a methodology that consistently makes most of the discipline's best gambits appear irrational, or only rational in light of external criteria, is certainly an inadequate methodology.

This Lakatosian approach to the question of economic methodology actually goes one step beyond Lakatos himself or any general supporters of the MSRP within the philosophy of natural science. After all, economics is a social science. Should social science methodologies be judged by the same standards as natural science methodologies? At first this may appear to be the old question of methodological monism—methodological monism being the position that the nature of social science is fundamentally the same as the nature of natural science, that science is science regardless of its domain of applicability.[9] On closer examination, though, we can see that the above proposal for appraising economic methodologies is not methodological monism at all, but what should be called "metamethodological monism." The argument is not that good economic theories are methodologically the same as good physical theories; the argument is that the metamethodological criteria used to judge the adequacy of a methodology should be the same whether that methodology concerns a social or a natural science. While these two positions need not be inconsistent, it is also the case that metamethodological monism does not necessarily imply methodological monism.

The rest of this discussion will simply presuppose metamethodological monism. Lakatos's metamethod for appraising philosophies of natural science will be applied to the MSRP as an economic methodology. If the discipline's "best" theories (or research programs) appear to be accepted for rational reasons, if the profession has supported "progressive" theories and eschewed "degenerating" ones via the MSRP criteria, then the MSRP should be positively appraised as an economic methodology. If not, then it should be appraised negatively.

PROGRESS AND THE MSRP

Lakatos calls a series of theories *theoretically progressive* if each new theory "has some excess empirical content over its predecessor, that is, if it predicts some novel, hitherto unexpected fact" (1970, p. 118). A series of theories is *empirically progressive* if "some of this excess empirical content is also corroborated." According to the MSRP, then, a research program is *progressive* if it meets the following two requirements. First, it must be *"consistently theoretically progressive,"* that is, "each successive link . . . predicts some new fact" (Lakatos, 1970, p. 134, italics added). And second, it must be at least *"intermittently empirically progressive,"* that is, "at least every now and then the increase in content should be seen to be retrospectively corroborated" (Lakatos, 1970, p. 134, italics added).

To understand what the two requirements of theoretical and empirical progressivity mean for the appraisal of the MSRP as an economic methodology it is necessary to carefully unpack Lakatos's notion of progress and the idea of increasing empirical content on which it depends. Most economists attempting to apply the MSRP to their own discipline have been too heavy-handed with respect to Lakatos's use of both the term *progress* and the concept of empirical content. This has led many economists to appraise economic theories as progressive when in fact no such progress occurred on the basis of Lakatos's criteria.

Unpacking these important concepts is facilitated by some intellectual genealogy. While there are many well-publicized differences between the MSRP and the falsificationism of Karl Popper, it should be remembered that Lakatos always considered his work to fall squarely within the Popperian tradition. This heritage is particularly apparent in both Lakatos's definition of empirical content and his notion of scientific progress.

For Karl Popper a good theory is one that is highly falsifiable. To be highly falsifiable the theory must make risky predictions, it must forbid certain things to happen, and the "more a theory forbids, the better it is" (Poper, 1965, p. 36). The *empirical content* (or empirical basis) of a theory is the set of all potential falsifiers; it is "the class of those observation statements, or basic statements, which *contradict* the theory" (Popper, 1965, p. 385). For a new theory to constitute *progress* in science, Popper contends that the new theory "must have new and testable consequences. . . . [I]t must lead to the prediction of phenomena which have not so far been observed" (Popper, 1965, p. 241). In addition to this increase in empirical content, the successful theory must also have *empirical success;* it must provide "a new success in predicting what had never been thought of before" (Popper, 1965, p. 243). Thus, for Popper, facts only count if they are "unknown," and "a fact which was already known before the theory's proposal does not support it" (Worrall, 1978, p. 46).

Since Lakatos requires empirical success only intermittently (every now and then) rather than consistently (at each step), his definition of progress is a bit less demanding than Popper's, but his definition of empirical content is identical to Popper's.[10] This means that, according to the MSRP, progress in science is *solely* dependent on increasing the Popperian empirical content of theories; that is, the *only* type of observation that can affect the appraisal of a theory is a *novel fact*. Explaining things already known is "cheap success" (Worrall, 1978, p. 49) and counts for nothing. Thus, the question of factual novelty is "probably more important to the Lakatosians than to any other philosophical school, since the former group think that *only* novel facts figure in a research program's appraisal" (Gardner, 1982, p. 2).

Members of the Lakatosian school quickly realized that the strict Popperian criterion of factual novelty used by Lakatos was too rigid. In work by Elie Zahar (1973) and John Worrall (1978), the notion of novelty was weakened to the following position. "The methodology of scientific research programs regards a theory as supported by any fact, a 'correct' description of which it implies, provided the fact was not used in the construction of the theory" (Worrall, 1978, p. 50).[11] Notice that, on this definition of novelty, the original Lakatosian (Popperian) requirement is a *sufficient* condition for novelty, but not *necessary;* that is, if the fact was not available it could not possibly be used in the construction of the theory, but just because it was not used in the construction of the theory does not mean that it was not available. As Worrall states, "This methodology embodies the simple rule that one can't use the same fact twice: once in the construction of a theory and then again in its support" (1978, P. 48).

This Worrall-Zahar notion of novelty is less strict than the original criterion offered by Lakatos; it allows more types of observational statements to count as novel. Since progress is solely determined by whether a research program predicts novel facts, this revision makes it easier for a scientific theory to be positively appraised on the basis of the MSRP. This should not be interpreted to mean that recent modifications to the MSRP have abandoned Lakatos's empirical criterion of progress. It is still true that even theoretical progress is an entirely empirical affair—not empirical in the sense of actually checking to see if certain observations are confirmed by the data, but empirical in the sense of predicting novel facts, predicting potentially falsifying observations not used in the construction of the theory. The revision introduced by Worrall-Zahar does not in any way change the notions of theoretical progress, empirical progress, or simply progress, as laid down by Lakatos. The only thing that has changed is the notion of novelty used in appraising these various forms of progress. It is no longer necessary for an observation to be unknown to count; it is only necessary that it not be used in the construction of the theory.

NOVEL FACTS IN ECONOMICS: KEYNES

How does the history of economic thought fare with respect to the MSRP? Most economic commentators on the subject have been cautiously sympathetic.[12] If in fact the revised notion of novelty discussed above has made a progressive appraisal even easier to obtain, should it not now be possible to internalize or rationalize, on the basis of the MSRP, most of what constitutes the discipline's major achievements?

The majority of the existing Lakatosian case studies in the history of economic thought would answer the above question affirmatively. The first of these positive appraisals was actually the first article formally to introduce Lakatos's work to the economics profession: Blaug (1975).[13] In his article, Blaug argued that the clearest case where a new research program appeared in the history of economic thought (and consequently the case the MSRP is most likely to appraise progressively) was the revolution in macroeconomic theory initiated by Keynes's *General Theory* in 1936. Blaug argues that "we see the Keynesian revolution as the replacement of a 'degenerating' research program by a 'progressive' one with 'excess empirical content' "(1976a, p. 164). Exactly how is it that the Keynesian program constituted this Lakatosian progress over its predecessor? We are told that the program's "principal novel prediction was the chronic tendency of competitive market economies to generate unemployment" (Blaug, 1976a, p 162). Are we therefore to believe that *unemployment* is the novel fact of *The General Theory*? Certainly, Blaug does not want to argue that unemployment was unknown at the time *The General Theory* was written (as the original Lakatosian criterion would require). No, definitely not; he states quite clearly, "The fact that there was unemployment in the 1930's was not itself in dispute" (Blaug, 1976a, p. 162)—a candid admission that unemployment could not be a novel fact in the original Lakatosian sense.

If unemployment fails the novelty test on the basis of Lakatos's original definition, could it pass on the basis of the Worrall-Zahar revised notion of novelty? Is it true that the concept of unemployment was not used in the construction of the theory? No, Keynes fails here as well. Blaug (1976a, pp. 162-63) makes it quite clear that *The General Theory* was written *precisely* to explain unemployment.[14] Unemployment was *the* fact *The General Theory* sought to explain, and the MSRP is quite clear that a theory is "not supported by the facts it was adjusted to fit" (Worrall, 1978, p. 51).

Even though unemployment does not qualify as a novel fact, what about the numerous potentially falsifiable auxiliary hypotheses introduced in *The General Theory?* Is it not true that the positive but less than unity relationship between changes in aggregate consumption and changes in national income (MPC), the inverse relationship between liquidity preference (LP) and the rate of interest, and the relationship between the marginal efficiency of capital (MEC) and the rate of interest were all actually Lakatosian novel facts that initiated at least twenty years of attempted corroboration? The answer to this question is both yes and no. Yes to the latter; it is true that almost an entire generation of economists wrestled with the empirical problems associated with the corroboration of these auxiliary hypotheses. No to the former; these are simply not novel facts by the MSRP criterion. The Keynesian concepts of MPC, LP, and MEC are all *used explicitly in the construction of the theory* and consequently are *not novel facts*. Keynes states clearly that "our ultimate independent variables as consisting . . . the three fundamental psychological factors, namely, the psychological propensity of consume, the psychological attitude to liquidity and the psychological expectation of future yield from capital assets" (1936, p. 247).

It appears that the most obvious candidates for novelty status are not in fact novel on the basis of the MSRP. Unemployment as well as these three fundamental psychological factors are used to construct the theory, and according to the MSRP "one can't use the same fact twice: once in the construction of a theory and then again in its support" (Worrall, 1978, p. 48). The claim that Keynesian economics won professional approval because it represented "progress" seems unjustified on the basis of a Lakatosian definition of progress.

This does not necessarily mean that the entire history of the Keynesian program was devoid of Lakatosian novel facts. For instance, it could be argued that the Phillips curve (Phillips, 1958) and the ensuing related literature represented a novel, and temporarily corroborated, fact for *The General Theory*. On the basis of *The General Theory* (particularly chapter 21), there is good reason to predict that such a relation would exist between unemployment and the rate of change in prices, but this empirical implication was *not* used in the construction of the theory, and therefore it might pass as a novel fact. The corroboration of the Phillips curve provided much support for the Keynesian program, and some would claim that its falsification contributed to the program's degeneracy during the 1970s. The problem is that even if the Phillips curve counts as a novel fact, it came late in the development of the program and thus, according to the MSRP, constituted only "intermittently theoretical progress" at best.

Finally, there is a fundamental problem with *any* attempt to explain the acceptance of Keynes's *General Theory* on the basis of a Popper-based notion of increasing empirical content. The problem is that by strict Popperian standards *The General Theory* had *less* empirical content than its predecessors, not more. This is easy to see if we remember that unemployment equilibria were explicitly forbidden by classical economics. *The General Theory* was indeed a more "general" theory; it explained how the economy could be in equilibrium at any level of employment. For Keynes, "the postulates of the classical theory are applicable to a special case only and not the general case, the situation which it assumes being a limiting point of the possible positions of equilibrium" (1936, p. 3). Since the Keynesian theory allowed for any level of employment, it was the classical theory that forbade more, and consequently had more empirical content. On strict Lakatosian criteria (at least with respect to employment), Keynes's theory "introduced no improvement since it did not forbid anything that had not been forbidden by the relevant theories he intended to improve upon" (Lakatos, 1970, p. 124).

There can be little doubt that Keynesian economics is a best gambit in economics if anything is. Regardless of one's opinion about the adequacy of Keynesian theory for analyzing today's economy, few, if any, economists would doubt its progressiveness in the 1930s. Certainly if Nobel prizewinning is indicative of the profession's best gambits, the articulation of the Keynesian program must be highly rated. On a strict application of the MSRP, such professional acceptance was *not* based on the progressiveness of the Keynesian program. If our only standards of rational scientific behavior in economics are those provided by the MSRP, the door is left open to the externalists, who patiently await. On the basis of Lakatos's metamethodology, the MSRP must receive negative marks with respect to Keynes.

NOVEL FACTS IN ECONOMICS:
GENERAL EQUILIBRIUM THEORY

In terms of the commitment of intellectual resources, there is probably not any aspect of economic theory (with the possible exception of Keynesian economics) that has received as much professional attention during the postwar era as the neo-Walrasian program of general equilibrium (GE). This research program has not only attracted a great number of economists; it has also attracted some of the profession's best analytical talent. Since the early 1950s the general equilibrium research program has been considered one of the most technically sophisticated research areas within economics, and it has dominated some of the profession's most prestigious journals. In spite of, and actually because of, all this professional commitment, general equilibrium theory probably stands as *the fundamental enigma* to any rationalizing methodology. The reason for this enigma is that on the basis of a strictly *falsificationist* criterion, general equilibrium theory has an abysmal record.[15]

E. Roy Weintraub applied the MSRP to the question of appraising GE in his 1979 book *Microfoundations*. In this work Weintraub argues that general equilibrium theory is the "hard core" of most current economic theory (1979, p. 37) and attempts to show, by way of an excellent history of the neo-Walrasian program, that sufficient progress has occurred in the derived propositions of the "protective belt" to provide a progressive appraisal of GE via the MSRP. In this view, any novel fact predicted by an economic theory based on the neoclassical framework of constrained optimization becomes a novel fact for GE. For instance, this allows for the implication that novel facts generated by recent applications of consumer-choice theory to new domains such as the economics of the family and human-capital theory constitute a progressive move for GE (Weintraub, 1979, p. 7).

The problem with this approach is that the MSRP requires at least theoretical progress *at each step* in the development of a research program. A positive appraisal of GE must demonstrate that each successful step in the program's history entailed an increase in empirical content. Therefore, the relevant questions to pose for a Lakatosian appraisal of GE are questions such as the following. Why were the original existence proofs for Walrasian equilibria produced in the early 1950s considered progressive? What novel facts were generated from the local and global stability theorems of the 1950s? And how was the empirical content of the program increased by the proof that the equilibrium price vector is unique when the weak axiom of revealed preference holds on aggregate excess demands?

While answers to such questions are necessary for a Lakatosian appraisal of GE, these are questions on which *Microfoundations* is silent. Weintraub provides an excellent survey of these theoretical topics, but never demonstrates how each of them (considered progressive by the relevant elite) generated the necessary novel facts. He simply packs all of modern economic theory into the neo-Walrasian portmanteau and then argues, "Alternatives to the neo-Walrasian program are, at present, nowhere near as complex, fruitful, and robust. The neo-Walrasian synthesis of microeconomics and macroeconomics, for all its defects, remains a serious object for study" (Weintraub, 1979, p. 68). While such statements may be true, they certainly do *not* show that the general equilibrium theory initiated by Leon Walras in 1874 and relegated to its esteemed position by the profession during the past thirty years has, at each step in its development, generated anything like the increased empirical content required for progressivity under the MSRP.[16] With GE, as with Keynesian economics, the MSRP has failed to rationalize the generally accepted best gambits of the profession, and Lakatos's metamethodology again demands a negative appraisal of the MSRP.

While the present article will not discuss specific alternatives to the MSRP, it is clear that any methodology that hopes to provide a strictly internal explanation of theory change in GE must provide an acceptable notion of *purely* theoretical progress. The MSRP's empirical criterion for theoretical progress simply will not do. It is not at all obvious that the strictly Walrasian program, with parametric prices, n-goods, and a *tâtonnement* adjustment mechanism, ever had *any* falsifiable empirical content, much less novel content at every step in its development. Yet, as Weintraub shows, and Schumpeter said many years ago, GE is "the basis of practically all the best work of our own time" (1954, p. 1026). Surely GE deserves more than the "degenerating since birth" appraisal it would receive on a strict application of the MSRP.

CONCLUSION

Keynesian economics and Walrasian general equilibrium theory are more clearly best gambits than any of the other economic research programs to which the MSRP has been applied. Therefore only a few comments will be made on these other applications. First, a few Lakatosian case studies provide a purely, or at least partially, negative appraisal of some specific program (or subprogram) in the history of economic thought.[17] Given the metamethodological task currently at hand, and the search for positive appraisals that task implies, these negative appraisals have little bearing on the issue at hand. Second, many of these case studies strain the term *novel fact* well past its breaking point. For instance, Rizzo (1982), which attempts to show progress in Austrian economics, has distorted "novel fact" beyond recognition. Thus, as with the negative appraisals, such work seems irrelevant for a general evaluation of the applicability of the MSRP to economics. Third and finally, there is the one exception that proves the rule. Neil de Marchi's (1976) case study of the Leontief paradox in fact demonstrates that it *is* possible to provide a legitimate Lakatosian rational reconstruction of a particular step in the development of an economic research program.[18] The problem is, one successful case study in a decade of such work only serves to demonstrate how ill-suited the MSRP actually is.

In summary, it has been demonstrated that, if "rationality" is restricted to the standards of the MSRP, then most of the history of economic thought (including the best gambits) has been irrational. Since rationalizing the best gambits is the expressed metamethodological test of the adequacy of the MSRP, the methodology must be negatively appraised. Of course, following Lakatos, a negative appraisal is no more grounds for the rejection of a methodology than it is for the rejection of a scientific theory. A methodology, like a scientific theory, should be abandoned only when it is degenerating *and* a more progressive alternative exists. The arguments presented above have demonstrated only the former.

Even though strict applications of the MSRP have proven inadequate, Lakatos's work can still provide valuable guidance in our search for a progressive methodological alternative (or alternatives).[19] First, just as Lakatos perceived the problem for the philosophy of natural science, economics needs a model of rational theory choice that will rationalize the great accomplishments in the history of economic thought without allowing just anything to pass itself off as progressive economic theory. Therefore, the *problem* facing those concerned with economic methodology is precisely the problem Lakatos faced for natural science. In fact, since the absence of falsificationist practice and the tendency toward external explanations of theory change are even greater in economics than in natural science, Lakatos's problems are not only present but actually *more* pronounced in economics.

Second, the failure of mechanical applications of Lakatos's philosophy of natural science does not necessarily preclude the success of a substantially modified version of the Lakatosian program.[20] A modified version of the MSRP, in order to be successful, must be written with the actual history of economic thought (at least the best gambits) squarely in sight.[21] It seems unlikely that these best gambits in the discipline's history can be accommodated as long as the Popperian notion of increasing empirical content is retained as a *necessary* condition for progress. Needless to say, rejection of Popper's notion of empirical content does not close the door on all empirical criteria of theory assessment. Lakatos chose Popperian fundamentalism on this point, but certainly "increasing the cardinality of the set of potential falsifiers" is not the only way that "the facts" can matter.

Whether future methodological alternatives are grown on Lakatosian roots or not, the Lakatosian episode in economic methodology has given us a number of important insights. First, Lakatosian philosophy of science has taught us to ask the right questions. Second, it has taught us to keep one eye on the actual history of economics as we attempt to answer those questions. And third, it has taught us to be very skeptical about uncritical borrowing from the philosophy of natural science.

NOTES

1. Worrall (1978) and Zahar (1973). For a comment on Lakatos's approval, see Worrall (1978, p. 65, n. 1).

2. This list of published work is not exhaustive, and there are a number of unpublished and forthcoming works on the topic.

3. Two comments are in order here. First, the reasons for being interested in applying Lakatos to economics may in fact be identical to the reasons for being interested in applying Lakatos to any other science. The question of the applicability to physical science is simply not addressed here; the discussion will concern *only economics*. Second, a similar question could be asked with respect to any other philosopher of science who has caught the eye of the economics profession (Karl Popper or Thomas Kuhn, for instance). While there is probably an interesting philosopher-specific story in each of these cases, such stories are again not addressed here; the discussion will concern *only Lakatos*, or at least only the Lakatosian school.

4. References on this point would include almost everything written about economic methodology in the past ten years, but the single best source is Blaug (1980a), a book whose entire message is ultimately "they don't practice the falsificationism they preach."

5. Lakatos's definition of "external" would include "economic conditions" as an external explanation of theory change. Obviously, then, what is external to physics is not external to economics. This is a point made initially by Axel Leijonhufvud (1976, p. 73).

6. The Lakatosian notion of progress will be discussed in detail below.

The defense of the MSRP on this issue is provided by Worrall (1978, pp. 62-63).

7. Lakatos weakens this a bit and argues that such rejection, like the rejection of a scientific theory, need not be absolute or swift. Regardless of this weakening, Lakatos never falters in expressing the view that the best methodology will render the "best gambits" rational. "Thus progress in the theory of scientific rationality is marked by discoveries of novel historical facts, by the reconstruction of a growing bulk of value-impregnated history as rational" (Lakatos, 1970, p. 118).

8. While this is not the place for a detailed discussion, it should be mentioned that there are several potential difficulties with this approach, the most obvious being the problem of circularity (see the discussion and references in Hands, 1984a). Several philosophers have severely criticized Lakatos on this point (see Kulka, 1977; and Laudan, 1977, pp. 155-70).

9. Methodological monism goes by many names; for instance, Spiro Latsis calls it "causalism" (1976b, p. 5). F. A. Hayek, extremely critical of methodological monism, calls it "scientism"—"a mechanical and uncritical application of habits of thought to fields different from those in which they have been formed" (1979, p. 24).

10. "The empirical basis of a theory is the set of its potential falsifiers: the set of those observational propositions which may disprove it" (Lakatos, 1970, p. 98, n. 2). The agreement with Popper on these points has become an important bone of contention for certain critics of Lakatos (see Laudan, 1977, p. 232, for instance).

11. Michael Gardner (1982) offers still another notion of factual novelty. According to Gardner, a fact is novel with respect to a particular theory if it was not known to the person who constructed the theory at the time he or she did so. Gardner claims this is an improvement over the Zahar-Worrall notion, which he calls "use-novelty." The jury is still out on this latest revision; currently the Worrall definition is considered to be the standard Lakatosian usage of "novel." One Lakatosian criticism of Gardner's suggestion might be that Gardner makes novelty a mental (or in Popper's terms, "World-2") entity, rather than an entity that is a *product* of the human mind (and thus part of Popper's "World-3").

12. Including this author (see Hands, 1979). Even economists with doubts about Lakatos have been somewhat restrained in their skepticism (see Archibald, 1979; Hutchison, 1976; Leijonhufvud, 1976; and Robbins, 1979). It should be noted that Leijonhufvud (1976, p. 79) actually suggested criticisms along the lines of those presented here, but failed to provide a detailed discussion.

13. Blaug's article is reprinted in Gutting (1980) and in slightly revised form in Latsis (1976a).

14. This is apparent from even the first two chapters of *The General Theory* (Keynes, 1936).

15. See Hands (1984a) for a list of references on this point. In fact, it is safe to say that *no* methodological defender of general equilibrium theory

even attempts to claim it is empirically falsifiable. See, for instance, Coddington (1975), Gibbard and Varian (1978), Hahn (1973), and Hausman (1981a).

16. In all fairness to Weintraub, *Microfoundations* (1979) was not the end of his interest in the question of the methodological appraisal of GE. In recent work he has turned to improving and extending Lakatos's work for this purpose, particularly by combining elements of Lakatos's philosophy of mathematics with the MSRP (see the suggestions to this effect in Weintraub, 1982). He has also improved the available history of the subject with his article on the history of the "existence" question (Weintraub, 1983).

17. Blaug (1976a, 1980b), Cross (1982), and Latsis (1976b).

18. And even this success is not without problems. For instance, claiming—as de Marchi (1976) seems to—that the Gale-Nikaido theorem on the global univalence of mappings constitutes a Lakatosian novel fact for the Ohlin-Samuelson research program is clearly not appropriate. Luckily this is not the only novel fact that de Marchi discusses, and the others actually seem to qualify. It also should be noted that, while this epoch in the history of international trade theory *can* be reconstructed along Lakatosian lines, this certainly is not the only convincing way to tell the story (see for example Chipman, 1966, pp. 44-57; and Takayama, 1972, pp. 75-91).

19. Certain economic methodologists now advocate methodological pluralism (Caldwell, 1982) or problem-dependent methodologies (Boland, 1982) rather than a single methodological position.

20. One such major revision of Lakatos by an economist is Remenyi (1979). Also see note 16 above on Weintraub.

21. See Hausman (1980).

5

Second Thoughts on "Second Thoughts": Reconsidering the Lakatosian Progress of *The General Theory*

This paper (Hands, 1990a) was originally prepared for the History of Economics Society meetings in Richmond, Virginia, in 1989 and it appeared in Review of Political Economy *in 1990. Although most of the chapters in this volume are arranged in chronological order, I felt that it was appropriate to move this relatively recent paper to Chapter 5 since it is a continuation of the "Second Thoughts" discussion from Chapter 4. The paper is a direct response to Ahonen (1989) and Blaug (1987), two papers critical of my "Second Thoughts" interpretation of the Keynesian revolution. It originally appeared with responses by Ahonen (1990) and Blaug (1990).*

The purpose of "Second Thoughts on Lakatos" (Hands, 1985a) was to apply Lakatosian standards to Imre Lakatos's work, to appraise Lakatos's methodology of scientific research programs (MSRP) by means of Lakatos's own metacriterion for evaluating scientific methodologies. According to Lakatos (1971a, 1978a), a general methodology or philosophy of science is to be appraised on the basis of how well it rationalizes the historical judgements of the scientific elite; an acceptable methodology "must reconstruct the acknowledgedly best gambits as 'scientific': if it fails to do so, it has to be rejected" (Lakatos, 1971a, p. 111).

Applying this Lakatosian metacriterion to economics, I argued that since Keynesian economics and Walrasian general equilibrium theory were clearly two of the profession's "best gambits," the MSRP could be appraised as an economic methodology by determining how "scientific" it rendered the history of these two research programs. Now since the MSRP (Lakatos, 1970) does not explicitly provide a demarcation criterion (a criterion for demarcating science from nonscience), my examination focused exclusively on the questions of the Lakatosian "progress" and/or "degeneracy" of the Keynesian and Walrasian programs. The argument was simply that "if the discipline's 'best' theories (or research programs) appear to be accepted for rational reasons, if the profession has supported 'progressive' theories and eschewed 'degenerating' ones via the MSRP criteria, then the MSRP should be positively appraised as an economic methodology. If not, then it should be appraised negatively" (Hands, 1985a, p. 5). This approach to appraising the MSRP presupposes a commitment to metamethodological monism that Lakatos probably would not have accepted, but I still think it represents a reasonable and straightforward way of judging Lakatos's methodology by his own standards.

The results of my search for Lakatosian progress in the Keynesian and Walrasian research programs were negative. According to Lakatos, the development of a particular research program is "progressive . . . if each new theory has some excess empirical content over its predecessor, that is, if it predicts some novel, hither to unexpected fact" (1970, p. 118). Since I found no such novel facts in either research program, I concluded, "that if 'rationality' is restricted to the standards of the MSRP, then most of the history of economic thought (including the best gambits) has been irrational. Since rationalizing the best gambits is the expressed meta-methodological test of the adequacy of the MSRP, the methodology must be negatively appraised" (1985a, p. 13). Of course, my conclusion did not imply that these two research programs were not "progressive"—to me, to the economics profession, or even on a more reasonable definition of "progress." My result was only that these two programs would not be deemed progressive on a strict application of Lakatosian standards. The problem is with the MSRP, not with the economics profession's best gambits.

In conducting the search for Lakatosian "novel facts," I felt (and still feel) that I was extremely generous to Lakatos. The original Lakatosian definition of novel facts was the same as Popper's standard for "independent testability"; that is, the theory "must have new and testable consequences (preferably consequences of a new kind); it must lead to the prediction of phenomena which have not so far been observed" (Popper, 1965, p. 243). This definition of novel facts—that the facts were "unknown" when the theory was proposed—is extremely restrictive; it would render "degenerate" much of the history of the best physical science. Given the restrictiveness of the Popperian definition, I followed many members of the Lakatosian school and adopted the more liberal (and thus more easy to satisfy) definition of novel facts proposed by John Worrall (1978) and Elie Zahar (1973). According to this neo-Lakatosian definition, a fact is novel if it was not actually "used in the construction of the theory" (Worrall, 1978, p. 50). "This methodology embodies the simple rule that one can't use the same fact twice: once in the construction of a theory and then again in its support" (Worrall, 1978, p. 48). The Worrall-Zahar definition of novelty is a "Lakatosian" definition, which (unlike the Popperian original) is liberal enough that it does not close the door on the search for "progress" before it even gets started. There are many different definitions of novel facts in the philosophical literature, and some of these I personally find much more satisfying than the neo-Lakatosian definition of Worrall and Zahar. I chose this definition only because it *is* now the *standard* Lakatosian definition and my purpose was to evaluate Lakatos's methodology by Lakatosian standards.

While E. Roy Weintraub's work on the neo-Walrasian research program (1985a, 1985b, 1988a) indirectly challenges my interpretation of Lakatosian progress within Walrasian economics, I am not aware of any work that *directly* challenges my "Second Thoughts" characterization of general equilibrium theory. Unfortunately, the same cannot be said for my evaluation of Keynesian economics. At least two authors, Guy Ahonen (1989) and Mark Blaug (1987), have directly criticized my evaluation of the (absence of) Lakatosian progress in Keynes's *General Theory* (1936). The purpose of this paper is to respond directly to the criticisms of these two authors. Ahonen and Blaug both argue that there was Lakatosian progress in Keynes's *General Theory,* and that based on this progressiveness my negative appraisal of the MSRP is unjustified. Because the arguments of the two authors and my responses to them are very different, I will examine each individually. I will first respond to Ahonen's criticisms and then my response to Blaug will follow in its own section. In the final section I will conclude and reflect on the entire discussion about Lakatos and economics.

RECONSIDERING LAKATOS: GUY AHONEN

Guy Ahonen (1989) raises three fundamental issues about my depiction of Lakatosian progress in Keynes's *General Theory.* All three of his comments refer to my reading of Lakatosian philosophy, rather than my reading of Keynesian economics. First, he argues that I have a "misconceived conception of falsification" (p. 257).[1] For Ahonen, Keynes's theory was more universal—and thus by Popper's standards more falsifiable—than classical economics, not less falsifiable as I argue.

Second, Ahonen (pp. 261-63) argues that my definition of novel facts is "arbitrary and misconceived" (p. 256), "somewhat strange," and "even arrogant" (p. 261). He claims that for Lakatos "novelty" is simply defined "as a theoretical implication that did not exist before" (p. 261). Third, Ahonen argues (pp. 264-65) that it does not make Lakatosian sense to talk about progress occurring from classical to Keynesian economics if these two theories have different "hard cores" and therefore belong to two entirely different research programs. Since I mostly agree with the last of these comments, I will start there and then work backward to the other two issues.

Ahonen's third comment is as follows: "Only if the hard core of two series of theories are identical can we actually speak of a 'progressive problemshift'" (p. 265). If Keynesian economics and classical economics have different hard cores, then "there cannot be discussion of increasing empirical content in Keynes's theory in relation to classical theory unless the theories are sufficiently similar" (p. 265).

On this point of Lakatosian philosophy Ahonen is actually correct. Lakatos discussed not only theoretical and empirical progress (the prediction and confirmation of novel facts), but heuristic progress ("non-ad hoc$_3$ness") as well.[2] "Lakatos considers a third type of progress, heuristic progress, which requires the changes to be consistent with the hard core of the program. His definitions of theoretical and empirical progress presuppose that conditions for this latter type of progress have been satisfied" (Hands, 1992a, p. 24). One program with one hard core may be progressive and a different program with a different hard core may be degenerating, but for Lakatos a "progressive problem shift" or "progress" from one theory to another is always within a particular program.

Now while Ahonen is correct about the intratheoretic nature of Lakatosian progress, I do not believe this amounts to a substantive criticism of my central thesis. I was certainly arguing in terms of a single research program (macroeconomics) and the Keynesian theory's being (or not being) a progressive problem shift within that program. Indeed it may be the case that the two programs do have different hard cores and therefore my discussion of a Lakatosian problem shift from one to the other is inappropriate. But this fact does not in any way support the MSRP as an economic methodology. It is a standard criticism of the MSRP that it provides no rational argument for choosing between two different research programs, that it presents only a vehicle for "monotheoretic" criticism.[3] Thus, if there are two programs, my discussion of progress would need to focus on the progress within each of these programs rather than on progress from one to the other; but this does not automatically lend support to the MSRP. This change only redirects our attention to a whole new set of problems with the MSRP.

Ahonen's second issue is a direct criticism of my use of "novel facts." While he agrees that for Lakatos novel facts are the sine qua non of scientific progress, he disagrees with my definition of novelty. After reproducing both of the Lakatosian definitions—the original Popperian definition that the fact had "never been thought of before," and the Worrall-Zahar definition that the fact was "not used in the construction of the theory"—Ahonen argues that "these definitions seem somewhat strange" (p. 261). He argues that I equate "Lakatosian factual novelty with observational surprise" (p. 261), finds this "totally misconceived" (p. 261), and then goes on to propose his own definition of novelty, which (not surprisingly) is easy to discover in *The General Theory*.

I fully agree with Ahonen's criticism of these definitions; I think that they *are* "somewhat strange" and that as the sole criterion for progress in science they are rather "misconceived." *That was precisely the point of my paper.* These *are* bad definitions of novelty, and we will not be able to make much sense of the history of economics or any other science if we are armed with only these concepts. The problem is, these are indeed the Lakatosian definitions of novelty, and according to the MSRP they are the *sole necessary condition for progress* in science. If one doubts these definitions and feels that other considerations should bear on what we judge as progressive (a view that Ahonen and I seem to share), then one simply rejects the MSRP. I have a difficult time understanding how someone can simultaneously hold that, on the one hand, the Lakatosian definition of novel facts is "misconceived" and, on the other, that the MSRP is the right tool for understanding progress in economics. One either believes that the history of Keynesian economics is replete with novel facts that were not used in the construction of the theory (as we will see Blaug does), or one does not believe that the Keynesian revolution constituted Lakatosian progress. Creating a new definition of novelty and then showing that Keynes constituted progress on this new definition says nothing about the MSRP; the result of such an inquiry may be independently interesting, but it has no bearing on the question my paper was designed to address.[4]

Ahonen's strongest criticism of my paper is the very first of his comments: that I presented "an arguable conception of falsification" (p. 257). I claimed that "there is a fundamental problem with *any* attempt to explain the acceptance of Keynes's *General Theory* on the basis of a Popper-based notion of increased empirical content. The problem is that by strict Popperian standards *The General Theory* had *less* empirical content than its predecessors, not more. . . . Since the Keynesian theory allowed for any level of employment, it was the classical theory which forbade more, and consequently had more empirical content" (Hands, 1985a, p. 10).

Ahonen claims precisely the opposite about the Popperian empirical content (and thus the falsifiability) of the two theories. Ahonen argues (pp. 257-60) that the Keynesian theory was more "universal," that it was a "generalization" of the classical theory, and that "Keynes refers to the extension of the field of phenomena which are to be explained, and thus results in an increase in the empirical content of economics" (p. 260).

To support his case, Ahonen reproduces (p. 5) the following schema from Popper's *Logic of Scientific Discovery:*

p: All *orbits of heavenly bodies* are *circles.*
q: All *orbits of planets* are *circles.*
r: All *orbits of heavenly bodies* are *ellipses.* (Popper, 1968, p. 122)

First consider only statements p and q. Notice that since the set of all planets is a proper subset of the set of all heavenly bodies, moving from statement q to statement p represents an increase in the "degree of universality" of the theory. Increased universality means an increase in the number of potential falsifiers and thus statement p is more falsifiable, has more Popperian empirical content, than statement q.

Ahonen likens the movement from classical to Keynesian economics to the movement from q to p: a movement that increases the degree of universality and thus the degree of falsifiability of macroeconomic theory. His argument is that, since Keynes assumed there were several different levels of employment and since classical theory assumed there was only one level (full employment), Keynesian theory provided an "extension of the field of phenomena" to be explained and thus resulted in "an increase in the empirical content of economics" (p. 260).

However convincing Ahonen's argument might seem on first reading, it cannot possibly be correct. Keynesian economics does not apply to a broader class of phenomena than classical theory; both Keynesian and classical theory are "about" developed capitalist economies. The difference is not in what the theories are about; the difference is in how the two theories characterize equilibrium: classical theory requires full employment for equilibrium; Keynesian theory does not.

Returning to Popper's schema (p, q, and r), the relevant analogy is not moving from q to p but rather moving from p to r. Notice that, in moving "from p to r, the *degree of precision* (of the predicate) decreases: circles are a proper subclass of ellipses; and if r is falsified, so is p, but not *vice versa*" (Popper, 1968, p. 122). Statement p is more precise and thus more falsifiable (has higher empirical content) than r.

To see how this notion of decreased precision applies to the case of Keynes and the classics let me introduce some symbolism. Let EE represent the economy's being in equilibrium, let N* represent the economy's being in full employment (N = N*), and let A represent aggregate demand being equal to aggregate supply—(D = Φ(N), in the language of *The General Theory*. Now, given this symbolism, we can write the principal conclusion of the classical theory (C) as

$$\text{EE implies N* and A.} \qquad \text{(C)}$$

Correspondingly, we can use this symbolism to write the principal conclusion of the Keynesian theory (K); it is

$$\text{EE implies A.} \qquad \text{(K)}$$

Notice that going from (C) to (K) is identical in form to going from p to r in Popper's schema.[5] Thus, moving from the classical theory to the Keynesian theory results in reduced precision; C is more testable than K. As I originally argued, "it was the classical theory which forbade more, and consequently had more empirical content" (Hands, 1985a, p. 10).

Again I would like to emphasize that my negative result is a reflection on Lakatos—in this case, on his inability to escape Popperian notions of empirical content—not a reflection on Keynes. I suspect that Keynesian macroeconomic theory was more testable than the classical macroeconomic theory it replaced—more testable and more corroborated. "Needless to say, rejection of Popper's notion of empirical content does not close the door on all empirical criteria of theory assessment. Lakatos chose Popperian fundamentalism on this point, but certainly 'increasing the cardinality of the set of potential falsifiers' is not the only way that 'the facts' can matter" (Hands, 1985a, p. 14).[6]

RECONSIDERING KEYNES: MARK BLAUG

Much of my discussion of Keynes in "Second Thoughts" was really a criticism of an earlier paper by Mark Blaug. Blaug (1976a) was one of the first papers to apply the MSRP explicitly to economics. In that paper it was argued that the Keynesian revolution constituted a shining example of Lakatosian progress.[7] Since my intent was to evaluate the Lakatosian progress of the economics profession's best gambits, examining Blaug's claim about Keynes was an obvious place to start. This makes my current reply actually a reply to a reply.

The first point of Blaug (1987) is that, even if one disregards the question of Lakatosian novel facts and even if one denies the "mythical picture of Keynes" as the lone voice of fiscal expansion (p. 7),[8] there is still much to recommend the Keynesian revolution. Keynes offered effective policy proposals that were contrary to the received view in British economic policy (though not necessarily American economic policy). Keynes advocated the "Great Heresy that an equilibrium level of income and output need not correspond to a situation of full employment" (p. 12), hinted that income redistribution might increase national income, and suggested that unguided capitalism "was doomed to secular stagnation" (p. 12). Such claims "proved irresistible to young economists, radicalized by years of depression" (p. 12). Keynes offered a theory that was easily operationalized, thus inviting the future swell of statistical analysis and applied econometrics (p. 14). He "achieved the optimum level of difficulty for intellectual success: not so simple as to be immediately accessible without personal effort and yet not so complex as virtually to defy comprehension" (p. 15). And finally, *The General Theory* was a work with an "efflorescence of ideas" (p. 14), an open-ended work that begged for further articulation and elaboration. Overall, *The General Theory* represented the right book with the right theory at the right time; as Blaug says in his penultimate sentence, "Keynes had caught a measure of substantive truth about the workings of an economic system that had not been vouchsafed to his predecessors" (p. 24).

Now, I would argue that for all of these reasons Keynes's *General Theory* constituted *progress,* but Blaug does not agree. For him progress occurs only as Lakatos defined it: predicting novel facts. This brings us to Blaug's second major point: that there were actually Lakatosian novel facts predicted by *The General Theory*. The principal novel fact was the "government spending multiplier," though there were others as well. It was a Lakatosian novel prediction of Keynesian economics "that the value of the instantaneous multiplier is greater than unity and that the more than proportionate impact of an increase in investment on income applies just as much to public as to private investment" (p. 16). This implied that "fiscal policy is capable, at least in principle, of raising real income up to the full employment ceiling within a single time-period" (p. 16). This novel fact, it is argued, was novel in both of the Lakatosian senses; it was both "unknown" (satisfying the original Lakatosian definition) and "not used in the construction of the theory" (per Worrall-Zahar). The government spending multiplier "was an unsuspected implication of the concept of the consumption function combined with the particular Keynesian definitions of savings and investment" (p. 17). Recent evidence discussed by Blaug (pp. 17-18) indicates that the early Keynesian estimates of the numerical value of the multiplier may not have been correct, but this only means there was not Lakatosian "empirical progress"; the novel prediction alone would be sufficient for Lakatosian "theoretical progress." Overall, Blaug's second point can be summarized as follows: "*The General Theory* gained adherents because Keynes made a novel prediction that seemed highly likely to be true" (p. 18).

This argument about the novelty of the multiplier seems correct to me.[9] As Blaug now concedes (p. 25, n. 1), unemployment certainly does not qualify as a Lakatosian novel fact (see Hands, 1985a); but as I am now willing to concede, the government spending multiplier does qualify. For Blaug this novelty has straightforward methodological implications; it "allows us to describe the Keynesian research programme as 'progressive' and that accounts for the rapid approval of Keynesian economics by the majority of the economics profession" (p. 23). While I will grant that the government spending multiplier is a legitimate novel fact (and maybe there are others as well), I still do not accept this novelty as the general *reason for the progress* of the Keynesian revolution, nor do I believe that it is appropriate to argue that such novelty "accounts for the rapid approval" of the profession. Let me address this second point first.

Lakatos did not argue that progressively reconstructing a particular episode in the history of science necessarily implied that this progress *was in fact the psychological reason for* the acceptance of the theory by the scientists of the day; "able to reconstruct progressively" does not necessarily imply "is actually the psychological reason for acceptance by the scientists involved." Throughout his discussion Blaug tends to suggest this connection between "positive appraisal" and "cause for acceptance." In addition to his claim that progress "accounts for the rapid approval" (p. 23), he also says that the multiplier "was decisive in converting the profession" (p. 19), and that the theory "gained adherence because" (p. 18) of novel facts. I find no such connection in Lakatos, and to do so would mix Popperian "World-2" entities (the mental activity of the individual scientist) with Popperian "World-3" entities (the objective appraisal of the theory)—something I doubt Lakatos would find acceptable. It seems perfectly consistent with Lakatos to argue that a particular theory was initially accepted for totally the wrong reasons and yet it actually generated novel facts and therefore should be appraised progressively with hindsight. In fact, I cannot really see any other way of reading Lakatos's discussion of "inductive" and "falsificationist" histories in Lakatos (1971a). I suspect that Blaug did not actually intend to forge this link between historical appraisal and psychological acceptance, but nonetheless the link does surface in Blaug (1987) and this is the way the argument seems to be interpreted by those reading the paper from a purely historical perspective.[10]

As for Blaug's other methodological implication—that the presence of Lakatosian novel facts "allows us to describe the Keynesian research programme as 'progressive' " (p. 23)—my opinion is basically unchanged from "Second Thoughts." Now that we have found Lakatosian novel facts, it *is* appropriate to anoint this particular best gambit with the title of "Lakatosian progress," but I still do not understand why we would want our best gambits so anointed. *Why would we want to accept the position that the sole necessary condition for scientific progress is predicting novel facts not used in the construction of the theory?* Surely humankind's greatest scientific accomplishments have amounted to more than this. We in economics and those in every other branch of science choose theories because they are deeper, simpler, more general, and/or more operational, explain known facts better, are more corroborated, are more consistent with what we consider to be more fundamental theories, and for many other reasons as well. Even if we can find a few novel facts here and there in the history of economics, and even if those novel facts seem to provide an occasional "clincher," the history of great economics is so much more than a list of these novel facts. Smith's invisible hand, the Ricardian law of comparative advantage, Walrasian notions of multimarket interdependency, Marshallian welfare economics, and the basic Keynesian notion of output and employment being determined by aggregate demand: these things constitute *great* economics; such theoretical developments have given us insight and scientific *progress*, not an occasional novel fact. The truly powerful insights in the history of economic thought—the things that motivated most of us to become economists—are lost in Lakatos's characterization. These important theoretical insights may be included in the hard core, but that's not what really matters for Lakatos; what really matters is predicting novel facts. Lakatos reduces the history of economic thought to the accumulation of a particular type of scientific artifact: novel facts. I do not understand why a historian of economic thought of Blaug's talent and stature would want to narrow our discussion in this way. Blaug starts his essay with a wonderfully convincing story about the rise of Keynesian ideas, an eminently interesting story and an eminently valuable story; then he turns to novel facts—a narrow definition of a narrow concept that produces a narrow view of science. He found some novel facts, so Keynes passed the Lakatosian test. I'm just not clear why we should care. In response to an earlier version of this Blaug paper, Bruce Caldwell stated "that Blaug is too good an intellectual historian to be anything more than a lousy Lakatosian" (1988b, p. 11). I too believe that Mark Blaug is a very good intellectual historian, and for that reason I think it would be a profound loss if he continued to become a better and better Lakatosian.

CONCLUSION AND REFLECTION

My view of Lakatosian economics has evolved over the years. At one point (Hands, 1979), I was rather enthusiastic; by Hands (1985a) I was quite negative. I continue to be negative, perhaps more so, about the MSRP as a general philosophy of science. I do not think that Lakatosian novel facts are either necessary or sufficient for scientific progress in economics or in any other science; novel facts may provide a clincher every now and then, like Halley's comet did for Newtonian physics, but they are nowhere near the whole story. Lakatos certainly had some important insights—the notion of a research *program*, the concepts of the positive and negative heuristics, the argument that theories are "born refuted," to name just a few—but he never really escaped the Popperian (and problematic) notions of empirical content and novelty. These latter problems and his unwillingness to move from appraisal to advice in a purportedly normative methodology seem to be the greatest vices of Lakatos's methodology.

On the other hand, historians of economic thought guided by Lakatosian ideas have produced some valuable studies. Blaug's work discussed above, as well as his original paper (1976a), are just two examples.[11] In particular I think the history of economic thought has been much better served by Lakatosian philosophy of science than by other inquiring frameworks of recent interest—Kuhn and rhetoric, for example. Despite the fact that novel predictions have been few and far between, hard cores and heuristics abound; and Lakatos's general link between empirical prediction and theoretical progress has helped initiate a serious historical examination of the role of empirical testing and econometrics in economics. Overall then I would still give the MSRP a negative appraisal as an economic methodology, but I would support the work of many of the historians of economic thought that have been guided by Lakatosian ideas. Maybe this is as it should be, for Lakatos himself stated,

> I hold that philosophy of science is more of a guide to the historian of science than to the scientist. Since I think that philosophies of rationality lag behind scientific rationality even today, I find it difficult fully to share Popper's optimism that a better philosophy of science will be of considerable help to the scientist. (Lakatos, 1978a, p. 154)

NOTES

1. All page numbers in this section refer to Ahonen (1989) unless otherwise indicated.

2. This is one of the main themes in Hands (1988).

3. See Koertge (1971) and Feyerabend (1975a, ch. 16), for example. Musgrave (1976) is an attempt to save Lakatos from this difficulty. The dilemma posed by Lakatos's unwillingness to pass intertheoretic (or interprogrammatic) judgement has led at least one recent commentator to conclude "that MSRP is beyond repair" (Anderson, 1986, p. 241).

4. In fairness to Ahonen it should be noted that at least one "Lakatosian" advocates a definition of novel facts similar to his "a theoretical implication that did not exist before" (Ahonen, 1989, p. 261) definition. Alan Musgrave defined novelty as "something which is not also predicted by its background theory" (1974, pp. 15-16). Musgrave actually attributes this definition to Lakatos—but it is Lakatos (1968), not Lakatos (1970) of the MSRP. The Musgrave definition is at best a minority view in the Lakatosian school: see Gardner (1982, p. 2, n. 1), Worrall (1978, p. 66, n. 8), and Carrier (1988).

5. Let EE = "orbits of heavenly bodies," let A = "has the equation $x^2/a^2 + y^2/b^2 = 1$," and let N* = "$a = b$."

6. I would also add that it is not even the case that rejecting Popper's notion of empirical content prevents one from being Popperian. Popper's concept of empirical content is so problematic that even neo-Popperians such as John Watkins have rejected it. Much of Watkins (1984) is an attempt to propose a neo-Popperian view of the growth of scientific knowledge that is not based on Popper's definition of empirical content.

7. This way of characterizing Blaug's problem situation may be unfair. What Blaug actually intended to do was to show that Lakatos's MSRP provided a "better framework for characterizing the nature of the Keynesian Revolution" (1987, p. 2) than the framework provided by Thomas Kuhn. His way of demonstrating this was to show that the Keynesian revolution constituted Lakatosian progress over classical economics. I must say that I am in complete agreement with Blaug's initial objective; Lakatos is "better" than Kuhn, and Blaug (1976a) makes a convincing argument for this. Of course,

my agreement with Blaug's initial objective does not imply that I agree with his general position on the MSRP. I am a "better" golfer than I was two years ago, but I am still not very good.

8. All page numbers in this section refer to Blaug (1987) unless otherwise indicated.

9. As pointed out in Caldwell (1988b, pp. 10-13), this does not mean that "falsificationism" was practiced (or should be) by the economics profession.

10. For example, G. K. Shaw (1988) interprets Blaug (1987) to say that this novel fact was the "reason for the rapid conversion to Keynesian economics" (Shaw, 1988, p. 143).

11. Many others are listed in Hands (1985a and 1992a).

6

Karl Popper and Economic Methodology:
A New Look

This paper (Hands, 1985b) appeared in the first issue of Economics and Philosophy. *The paper argues that there are fundamental differences between Popper$_n$, the falsificationist philosopher of natural science, and Popper$_s$, the philosopher of social science who advocates the method of situational analysis. In preparing this chapter I considered changing Popper$_n$ (for "natural" science) to Popper$_f$ (for "falsificationist") since the latter seems to convey more accurately the desired distinction. In the end, though, I left the subscripts as they originally appeared since these labels have, rightly or wrongly, made their way into the later literature.*

Karl Popper's falsificationist philosophy of science is frequently discussed in the recent literature on economic methodology and there seems to be two basic points of agreement about his work. First, most economists take Popper's falsificationist method of bold conjecture and severe test to be the correct characterization of scientific conduct in the physical sciences. Second, most economists admit that economic theory fails miserably when judged by these same falsificationist standards. As Spiro Latsis states, "the development of economic analysis would look a dismal affair through falsificationist spectacles" (1976b, p. 8).

Since most economic methodologists agree that economics has failed to live up to falsificationist standards, the disagreement is directed toward the question of whether falsificationism *should* be practiced. Empirical hard-liners such as Terence Hutchison consider strict adherence to Popperian falsificationism to be absolutely essential for the growth of economic knowledge. For Hutchison, falsificationism defends the economic house against an invasion by ideologues and those who traffic in "complacent, pretentious, and noxious dogmatism" (1976, p. 203). At the other extreme, "open door" methodologists such as Donald McCloskey (1983) view falsificationism as simply a variant of "modernist" philosophy—a view that should be replaced by the candid admission that economic discourse is fundamentally rhetorical.

Although many different views regarding the applicability of Popper's falsificationism have been offered, one thing is strikingly absent from these presentations. There is seldom any reference to what Popper has actually written about economics. The Popper who is presented by economists is Popper the philosopher of *natural* science. The works most often cited are *The Logic of Scientific Discovery* (1968; 1st ed., 1959), and various summaries of the same position in *Conjectures and Refutations* (1965). In these works, Popper is primarily concerned with "demarcation," with formulating a criterion for distinguishing natural science from metaphysics. It is in these works that the basic falsificationist program of bold conjecture and severe test is delineated, and these works are the most strictly falsificationist of all Popper's writings.[1]

The problem with this strict falsificationist view of Popper is that it is inconsistent with what Popper and Popperians within philosophy of science have *actually written about economics* and the other social sciences. In the few places where Popper directly refers to economics, he is almost never discussing his falsificationist approach to natural science. Instead, economics is discussed in the context of his "situational analysis" or "situational logic" approach to historical and social explanation. Of course, if situational analysis were entirely consistent with Popper's falsificationist philosophy of social science, then the current characterization of Popper by economic methodologists would still be appropriate. Unfortunately, though, this is not the case. Most philosophers who have addressed this issue, including Popper himself, have suggested that situational analysis produces explanations that are actually unique to social science and would be less than adequate when judged by strict falsificationist standards.

The possibility of Popper's having a nonfalsificationist view of economic method raises a number of interesting questions. Exactly what is the relationship between economics and situational analysis? Will a detailed study of situational analysis provide additional insights into the methodological questions of economics that would be unavailable through falsificationist spectacles? What questions does such a potential dualism raise regarding methodological monism—the view ostensibly supported by Popper, that the method of social science and the method of natural science should not differ in significant ways? And finally, what does Popper really advise about practicing the science of economics?

This paper will address these and related questions. In the first section, situational analysis will be defined and discussed in detail. In the second section, the relationship between situational analysis and the literature on economic methodology will be examined with particular emphasis on the so-called rationality principle. These first two sections will use two of Popper's papers that have regrettably received little attention from economists: his 1967 French paper on economics, and his debate (Popper, 1976a) with the Frankfurt school. The third section will discuss the problem of theory choice in situational analysis with particular emphasis on questions of economic methodology. The final section contains suggestions for future research.

SITUATIONAL ANALYSIS AND THE
RATIONALITY PRINCIPLE

The first problem one encounters in trying to unpack Popper's notion of situational analysis is that Popper's own presentations of his view are often vague and seemingly inconsistent. These inconsistencies seem to emerge both within particular presentations and between alternative presentations. This difficulty has been noted by critics and defenders alike. Even I. C. Jarvie, a staunch supporter of Popper's approach, admits that Popper's view is "nowhere fully explained outside of lectures" (1972, p. 5). Critics, such as Latsis (1983), go so far as to accuse Popper of making an "obscure and unsatisfactory" (p. 132) presentation that "seems either confused or deliberately elusive" (p. 133). This confusion may be one of the reasons why economists have shied away from Popper's explanation of situational analysis and focused on falsificationism where even critics agree that his position is more clear and straightforward.

A brief bit of exegesis from Popper's intellectual autobiography (1976b) will help demonstrate how potentially confusing Popper's presentation of situational analysis can be. In describing the filiation of his ideas on situational logic, Popper explains, "The main point here was an attempt *to generalize the method of economic theory (marginal utility theory) so as to become applicable to the other theoretical social sciences*" (p. 117-18).[2] Later, in discussing evolutionary biology, Popper states that he also considers Darwinism to be an "application" of what he calls "situational logic" (Popper, 1976b, p. 168). From these two quotes one might conclude that Popper believes the rather strange proposition that evolutionary biology is merely an application of the method of marginal utility theory. If this is not sufficiently confusing, one has only to read on a bit more to find Popper saying that, once we accept Darwinian theory as situational logic, then we can "explain the strange similarity between my theory of the growth of knowledge and Darwinism: both would be cases of situational logic" (Popper, 1976b, p. 169). This allows us to conclude, further, that Popper's entire philosophy of science is simply an application of the method of neoclassical economic theory. The possibility of such a seemingly incredible conclusion only exemplifies the problems that await anyone trying to interpret Popper's writings on this topic.

In spite of these problems, if one is not too demanding, it *is* possible to distill from Popper a reasonably consistent characterization of what he means by situational analysis. The fundamental tenet of the program is that we should seek explanations of social behavior in terms of the "situation" in which the agents find themselves. Given this objective situation, there will be a unique action that follows naturally from the "logic" of the situation. The observed action is then explained as a "rational" or "logical" response to the objective situational environment in which the persons found themselves. Such explanations must be animated by a "rationality principle" stating simply that people act in a way that is appropriate to their situation. "Thus to explain a particular action, we describe the agent's situation, i.e., his or her goals and beliefs determine which action is appropriate to the situation as the agent perceives it, and with the addition of the Rationality Principle, deduce what the agent will (or did) do" (Koertge, 1974, p. 201).[3]

The use of the rationality principle in social science explanations raises a number of questions. For one thing, there is the question of necessity. Is the rationality principle necessary for explanations in social sciences or merely one possible approach? Popper himself seems ambivalent on this point. In early presentations (1961, 1966), Popper seemed to characterize situational analysis as merely one possible approach to explanation in social science. Later (1967, 1976a), he seems to posit it as the only approach to explanation in the social sciences. Most commentators have presented this latter view as the true "Popperian" position regarding necessity. Noretta Koertge states explicitly "Popper claims that the *only* means that we possess for explaining and understanding social events is situational logic used in conjunction with the Rationality Principle" (1974, p. 199). Similarly, John Watkins (1970, p. 167) argues that the rationality principle is necessary, at least for historical explanations; while Spiro Latsis (1983, p. 136) argues it is "indispensable" for explanations of social behavior. This necessity certainly seems paradoxical given Popper's stated commitment to methodological monism.[4] How can social science be like natural science when the only method available to social science is unavailable in the natural sciences?

Another important question pertains to the falsifiability of the rationality principle. Since the rationality principle serves as a *general law* in the explicans of social science explanations, it must satisfy the necessary conditions for general law status, the most important of these being that it be falsifiable. If it is not (at least potentially) falsifiable, then nothing demarcates explanations that employ the rationality principle from merely metaphysical explanations.

On the question of falsifiability it again appears that Popper has changed his mind during the course of his career. In his early discussion (1961), Popper emphasized the similarities of the natural and social sciences, particularly emphasizing that falsificationism is the proper approach to both. Ostensibly, he intended this falsificationist practice to extend to social sciences that employ the rationality principle. In more recent work (particularly 1967, 1976a), Popper has taken a different and seemingly paradoxical position. He now argues that the rationality principle is *false* and yet it is *not* testable (i.e., unfalsifiable). How can something be false and yet unfalsifiable? The answer is that it can be false as a *universal* principle and yet unfalsifiable in any *particular* application. Let us consider these points separately.

Regarding the falsity of the rationality principle, Popper simply means that it is not always true. Since this principle is stated as a universal law—"*all* agents act appropriately"—it is false if there exists a single agent who violates it. Popper argues that, though such agents exist (i.e., it is false), the principle is approximately true; that is, *most but not all* agents obey the principle (1967, pp. 144-46; 1976a, p. 103).

Regarding the unfalsifiability of the principle, Popper actually offers two separate arguments. The first depends exclusively on his claim that it is necessary for social science. Since the rationality principle constitutes a general law necessary to animate *every* theory in social science, the falsification of a *specific* theory does not falsify the rationality principle. The falsification of a specific theory only means that we have misspecified the situation; that is, we have attributed the wrong preferences or constraints to the individual.

As an economic example of Popper's first argument, consider a worker confronted with a higher real wage. One falsifiable prediction might be that the individual will work more hours per week: the substitution effect outweighs the income effect. Another falsifiable prediction might be that the individual will work fewer hours per week: the income effect outweighs the substitution effect. Both of these testable hypotheses entail the rationality principle; in both cases the individual is behaving appropriately: choosing his or her most preferred bundle, subject to given constraints. Now suppose that we observe a higher real wage followed by an increase in the amount of labor supplied per week; that is, the latter hypothesis is falsified. Does this mean that the worker acted inappropriately, that he or she was irrational? Clearly not. Since the rationality principle is contained in both hypotheses, it is not indicted by the failure of either one.

Popper (1967) also provides a second (and perhaps more convincing) defense for the unfalsifiability of the rationality principle. This second argument is the one supported by philosophical commentators such as Farr (1983), Koertge (1974, 1975, 1979a), and Watkins (1970). The argument is that the rationality principle *is* potentially falsifiable. That is, we *could* choose to reject it; we simply *decide* not to. We make a methodological decision that, when faced with a falsifying observation about a particular agent, we will *"cling to the rationality principle and revise our hypothesis about his aims and beliefs"* (Watkins, 1970, p. 173). This makes the "decision" to protect the rationality principle a matter of methodological choice, similar to Popper's falsificationist methodological choice always to accept anomalous observations as unproblematic and blame the theory. The justification for this protective decision varies among authors. For Watkins, the rationality principle must be "treated as unfalsifiable in the interest of the falsifiability of the whole system" (1970, p. 173). On the other hand, for Koertge (1979a, p. 93) the rationality principle is simply the Lakatosian "hard core" of the Popperian research program in the social sciences. Since, according to Lakatos (1970), the negative heuristic forbids us to direct criticism at the hard core, the rationality principle can be safely insulated in this way. A third view, related to the second, is that deciding to hold onto the rationality principle is in effect a decision to "do economics" as opposed to "doing" some other social science that might explain the same behavior.

From the perspective of Popper's demarcation criterion, it does not matter whether the rationality principle is protected as the result of a methodological decision, or because the nature of explanations in the social sciences precludes it. The fact remains that the rationality principle is not actually subjected to the severe testing necessary to demarcate scientific theories from metaphysical theories. Thus, by strictly Popperian standards, explanations that include the rationality principle (necessary or not) are as close to metaphysical explanations as they are to scientific explanations. As Koertge (1979a, p. 93) has suggested, this conclusion discredits Popper's use of his demarcation criterion to criticize the scientific standing of Marxian and Freudian theories. These theories can hardly be criticized for failing to provide scientific explanations if no other social science provides such explanations either.[5] At the very least it appears that Popper's writings on social science suggest a type of scientific behavior that would fail the acid test set forth by Popper the falsificationist. Popper's *claim* that the social and natural sciences are fundamentally the same seems contradicted by the very social science he advocates.

SITUATIONAL ANALYSIS AND ECONOMIC METHODOLOGY

There is no doubt that Popper had economics squarely in mind when he initially developed the situational analysis approach to social science. "It is, in fact, the method of economic analysis" (Popper, 1966, p. 97).[6] The rationality principle that Popper requires to translate beliefs and desires into action is simply a more general version of the "rationality postulate" or "maximization hypothesis" that has been a perennial bone of contention throughout the history of economic thought.[7] Yet, despite this obvious relationship, Popper's own writings on situational analysis as well as the various philosophical commentaries on his writings have not been given adequate treatment by economic methodologists.

For example, in Bruce Caldwell's otherwise masterful survey (1982), he asks how a "Popperian" would react in the face of a disconfirming instance of a theory that included the rationality postulate. He answers that a Popperian

> would insist that the data be clean, the test design straightforward, and initial conditions checked, all prior to the test. Such procedures guarantee that the test is a test of the rationality hypothesis only, and not a test of the other components. A Popperian would further insist that the hypothesis be either rejected or modified in the face of disconfirming instances. (Caldwell, 1982, p. 155)

While Caldwell is certainly correct that such behavior would characterize a falsificationist, "Popper emphasizes that regardless of how irrational a given action may appear to be, we should never conclude that the RP[8] is false (although he admits that it may be); rather we should always assume that we have not accurately described the agent's perceived situation" (Koertge, 1974, p. 201).

Thus while Caldwell's statement is correct with respect to Popper the falsificationist, it is not correct for Popper the philosopher of social science. In order to separate these two Poppers in the discussion that follows, I will follow the much-maligned Lakatosian tradition of subscripting Poppers.[9] Let Popper$_s$ be the Popper who has written on situational analysis and the rationality principle. The position of Popper$_s$ was described in the preceding section and is best represented by Popper (1967 and 1976a). The other Popper, the well-known falsificationist, will be referred to as Popper$_n$ (for natural science).[10] Popper$_n$ is the standard Popper, the advocate of bold conjecture and severe test. Popper$_n$ is probably best represented by Popper (1965 and 1968).

While economic methodologists such as Caldwell can only be accused of the minor infraction of *neglecting* Popper$_s$,[11] others—particularly Terence Hutchison—might be more harshly accused. Hutchison has long been the most consistent "Popperian" among methodologically inclined economists. He has consistently employed the work of Popper$_n$ to advocate a more falsificationist approach to economic method. In a recent work on economic methodology (1981), Hutchison presents a detailed defense of falsificationist practice and even lists at least three reasons why falsificationism is *even more* important for economics than it is for branches of natural science. Only once in this work does Hutchison make reference to Popper's writings on the rationality principle. In a footnote, he translates from Popper's 1967 French paper on the topic: *"The rationality principle is false. There seems to me to be no way of escaping this conclusion. . . . It should be considered as belonging to empirical theory, and should be submitted to tests"* (quoted in Hutchison, 1981, p. 303, n. 17).[12]

At best, Hutchison's quote presents us with a confused picture of Popper$_s$. For one thing, in Popper's original paper, the two parts of Hutchison's quote are separated by two important paragraphs of text. For another, in the section following the last line of Hutchison's quote, Popper presents a series of arguments concluding, that even though the principle *is* testable, it should *not* be rejected. Popper's argument on this point has already been presented above as the second of his reasons for protecting the rationality principle from falsification: that it is simply "good methodological practice" to do so. Since the rationality principle is false, Popper argues that rejecting it tells us nothing new. On the other hand, the characterization of the situation may be either true or false and thus there is more to "learn" from a reexamination of the situation. This is hardly the position one would attribute to Popper solely on the basis of Hutchison's quote. Actually, this possible misrepresentation of Popper$_s$ does not matter to Hutchison's overall presentation, since Popper$_s$ is only mentioned this one time in Hutchison's paper. The rest of the discussion is strictly falsificationist, with many pleas for the "actual testing of the assumption" (Hutchison, 1981, p. 293) of rationality without any reference to Popper's or other philosophers' arguments why that might be impossible or at least methodologically unsound.

The only detailed discussion of economic methodology that does discuss Popper$_s$ is Latsis (1976b).[13] This paper, fairly well known for its application of Lakatos's methodology of scientific research programs to economics, presents three approaches to economic method (other than Lakatos's): apriorism, falsificationism, and conventionalism (Latsis, 1976b, pp. 3-14). One might suppose, following economic practice, that Popper's name would only appear in reference to falsificationism, but this is not the case in Latsis's discussion. Falsificationism is attributed primarily to Hutchison, with only one footnote reference to Popper in the section where the falsificationist position is explained (Latsis, 1976b, pp. 7-9). The primary reference to Popper is in the section on "apriorism." Popper's rationality principle is associated with the Austrian apriorism of Ludwig von Mises and to a lesser extent F. A. Hayek. According to Latsis the "rationalistic approach" (1976b, p. 4) of Popper and the Austrians has produced "situational determinism which has been the dominant research program of neoclassical microeconomic theory" (p. 16). This situational determinist program is contrasted sharply with falsificationism by Latsis, thus providing the only published comparison of Popper$_s$ on economics with Popper$_n$ on economics.

Space considerations preclude a detailed discussion of the entire Latsis (1976b) paper, but at least one comment should be made. Latsis's inclusion of Popper$_s$ into the Austrian apriorist camp probably *overstates* the differences between Popper$_s$ and Popper$_n$. It is not at all clear that Hayek should be included in the Austrian apriorist camp, and Popper (either one) is certainly farther away from the Austrian position than Hayek. While the question of the relationship between Hayek's methodological position and that of his Austrian mentors, as well as the question of the relationship between Popper$_s$ and Hayek, are both fertile topics for future research,[14] Latsis's assertion that they are all three roughly equivalent seems entirely unjustified. The evidence of future research will probably demonstrate that Popper$_s$ is as much misrepresented by Latsis's apriorist characterization as he has been by those who try to reduce him to Popper$_n$.

It appears that those writing on economic methodology should not receive very high marks for their treatment of situational analysis. Most simply ignore Popper$_s$ altogether and focus exclusively on Popper$_n$. The few economists who do refer to Popper$_s$ seem less than certain about his identity. He is either subsumed under Popper$_n$ or characterized as the opposite extreme: an Austrian apriorist. On the other hand, philosophers writing about situational analysis should not receive very high marks for their treatment of economics, either. Though Popper is quite clear regarding the economic lineage of situational analysis, most philosophers presenting the topic tend to focus their attention on applications from other social sciences. In the few places where philosophers do discuss the relationship between situational analysis and economics, their characterization of economics is less than accurate. Some of these issues will be discussed in the following section.

THEORY CHANGE AND SITUATIONAL ANALYSIS

What are the methodological implications of the arguments in the two preceding sections? If Popper's advice on social science is different from his advice on natural science, and if most economic methodologists have failed to recognize this difference, what can be learned about the proper (even Popperian) approach to economic methodology? What can economists learn from Popperian philosophy of social science, and vice versa?

In many respects the answers to these questions are negative. Philosophers who have worked on Popper's approach to situational analysis seem to have stumbled over precisely the same difficulties as economists writing about methodology. For instance, the discussion of the falsifiability of the rationality principle by philosophers (presented under the first heading, above) parallels the discussion of the falsifiability of the maximization hypothesis by economists. Since the rationality principle is used in scientific explanations, most sympathetic philosophers would like to say that it is falsifiable and that it is accepted because of its ability to survive severe empirical tests. In fact, it is almost never tested and is known to be false as a general law. Yet most philosophers who have examined the topic conclude that it plays an important role in providing "explanations" that are nontautological, guide us to novel discoveries, and tend to be more acceptable than those offered without the rationality principle. At the very least, the rationality principle seems to be "confirmable and influential metaphysics" (Koertge, 1974, p. 201). A similar position has recently been taken by the economist Lawrence Boland (1981) with respect to the maximization hypothesis in economics. Boland found the hypothesis to be a nonfalsifiable and yet useful and nontautological metaphysical presupposition of the neoclassical research program.

One thing that economists should reasonably expect to gain from philosophers of science discussing $Popper_s$ is a criterion for progressive theory change for theories involving the rationality principle. Most philosophies of natural science provide such a criterion, and certainly Popper's writings on falsificationism provide such a standard. $Popper_n$ tells us to confront the theory with the facts; attempt to falsify it; and if it is falsified, reject it. If we have a number of theories none of which have been falsified, $Popper_n$ tells us to preserve the one with the most empirical content, the one that makes the boldest conjectures, the one that sticks its neck out the most. As Ernest Gellner elegantly describes the $Popper_n$ position: "The advancement of knowledge is attained, on this view, by seeking formulations which offer the enemy, namely nature, the most exposed and extended surface" (1974, p. 171). Even if two nonfalsified theories have equal empirical content, $Popper_n$ still provides us with a criterion for choice. We should choose the theory that predicts *novel* facts. The preferred theory should provide "a new success in predicting what had never been thought of before" (Popper, 1965, p. 243).

$Popper_n$ is certainly not alone in his concern over establishing a criterion for theory choice. It is safe to say that *most* philosophy of science is concerned with precisely this question and its corollaries. When should we accept a scientific theory? When should we reject a scientific theory? How can we make "progressive" changes in a theoretical system found to be problematic? Even a philosopher like Paul Feyerabend (1975a), who concludes quite negatively that there are no rational standards for theory choice, still retains theory choice as the main topic of discussion for the philosophy of science.

Unlike $Popper_n$, $Popper_s$ (and the principal philosopher commentators on $Popper_s$) has very little to offer regarding theory choice. Little or no guidance is offered in response to the important questions of philosophy of science. When should we accept or reject an explanatory theory involving the rationality principle? How do we tell a good situational analysis from a bad situational analysis? How should we make "progressive" modifications to a theoretical system that contains the rationality principle? Most of the literature advocating the use of situational analysis—Jarvie (1972) being an excellent example—concentrates on providing "successful" examples of the method. The intent of these studies is to demonstrate by example that situational analysis is applicable to social sciences other than economics. While such work may be methodologically consoling to the economics profession, it offers very little help with the hard questions of economic theory choice. For the philosophical literature to provide guidance to economists, their emphasis must move away from the general advocacy of situational analysis as a research program in social science to the specific details of successful decision making within the program. An economic example will help clarify this point.

In his highly regarded history of utility theory (1965), George Stigler argues that one of the major advances in utility theory came from Edgeworth's "generalization" of the utility function. Prior to Edgeworth, major neoclassical theorists (including Jevons, Marshall, Menger, and Walras) had assumed all utility functions to be additively separable. That is, if n goods $(X_1, X_2, \ldots, X_n)$ are being consumed, then the utility function has the form $U_1(X_1) + U_2(X_2) + \ldots + U_n(X_n)$. This form obviously rules out the case where increased consumption of one good increases or decreases the marginal utility of any other. In addition, and more important for empirical testing, this additive separability implies[15] that all goods are "normal" and therefore all demand curves are negatively sloped. Empirically it seems that all demand curves are *not* negatively sloped, and Stigler argues that this falsification is a primary reason for the eventual acceptance of the more general form that allows for "Giffen" goods. Stigler (and the majority of the profession) considers this change from additive to generalized utility functions to be a "progressive" theory within neoclassical demand theory.

From the perspective of Popper$_s$, the change to a generalized utility function is an example of a theory change *within* an explanatory theory containing the rationality principle. When the earlier additive form was confronted with the fact of upward sloping demand, the response was *not* to abandon the rationality principle—or maximization hypothesis—as an explanation of consumer behavior. Prima facie, the profession's response was exactly as Popper$_s$ would recommend: hold onto the rationality principle and revise the hypothesis about the agent's aim and beliefs.[16] The problem is that neither Popper$_s$ nor philosophers discussing this work explain *how* we should revise our characterization of the agent's beliefs. The choice made by the economics profession was to broaden the class of utility functions so as to protect the theory from falsification. This adjustment may or may not be progressive with respect to Popper$_s$ since neither he nor his advocates supply a criterion for such change, but this adjustment would definitely *not* be considered progressive for Popper$_n$. Such a modification is ad hoc; it modifies the theory in a way that renders it *less falsifiable*. Additive utility functions constitute a bolder conjecture than utility functions of an unspecified form, because the generalized form has a smaller set of potential falsifiers. Therefore, a change to the general form constitutes a clear case of an ad hoc and thus degenerative modification of the theory.[17]

As a casual examination of the history of thought reveals, modifications (such as the change to generalized utility functions) that reduce the empirical content of the theory are often considered, ex post facto, to be progressive moves by the majority of economists. The existing philosophical literature gives no indication whether such professional behavior should be applauded or condemned by Popperian social scientists. Certainly $Popper_n$ would condemn such moves—but if $Popper_n$ and $Popper_s$ are indeed different, why should we assume that the same rules of theory choice apply to Popperian social science as to Popperian natural science? Although this question is not properly addressed in the existing literature on Popperian philosophy of social science, it remains an important question. This and related topics are discussed in the concluding section.

CONCLUSION

For philosophers of social science (particularly Popperian), the above discussion suggests at least two directions in which research should be extended, both of which concern the question of theory choice in situational analysis. First, a closer examination of Popper's writings on metaphysics seems to be in order. If $Popper_s$ has a larger metaphysical component than $Popper_n$, then maybe Popper's few suggestions about the dynamics of metaphysical research programs[18] should be considered in proposing rules for theory choice in social science. Since Popper has argued that situational analysis and evolutionary biology are closely related, this investigation might also provide insights into the philosophy of biology.

Second, and perhaps more importantly, a detailed examination of theory choice *within economics* (particularly demand theory) needs to be undertaken by Popperian philosophers of science. Given the historical bent of recent philosophy of science and since Popper considered economics to be *the* model for situational analysis, an examination of theory choice within economics seems to be the obvious place to start an investigation of theory choice within situational analysis. It is possible, of course, that such an examination will prove damaging to the general method of situational analysis. If philosophers fail to uncover any type of "Popperian progress" in economics, then the entire Popperian approach to social science might be indicted. After all, Popper is quite clear that situational analysis is the method of economics (particularly neoclassical demand theory); failure within its indigenous domain would certainly reflect badly on the entire program. Not only is the success of Popperian social science related to the success of economics, given Popper's stated motivations for his demarcation criterion;[19] economics reflects, either positively or negatively, on the entire Popperian approach to knowledge. Given this potential importance, and given the general Popperian maxim to "look for trouble," it seems that a serious study of the history of economic thought by a Popperian philosopher is long overdue.

The above discussion also has a number of implications for methodologically inclined economists. First, the well-known fact that economic theory fails to live up to the falsificationist standards of $Popper_n$ need not imply that economics deserves the methodological contempt that it has recently received from economists sympathetic to Popper. It is apparent that even Popper himself recognizes strict falsificationism to be an extremely problematic program for the social sciences. It is therefore entirely possible to remain pristinely Popperian without finding the entire history of economic thought to be a gross violation of the scientific method. Of course, even after a serious examination of $Popper_s$, some Popperian economists may still wish to argue that strict adherence to the falsificationist rules of $Popper_n$ constitutes the only correct approach to economic inquiry; but that is a different matter than simply neglecting $Popper_s$.

Second, there is a lesson for economic methodologists who are generally not of the Popperian persuasion. Authors who have abandoned Popper simply because falsificationism is too inflexible to serve as a methodological guide for economics may need to reexamine the work of $Popper_s$. The neglect of $Popper_s$ may have led to a methodological mistreatment of Popper and Popperian philosophy by economists.

Third and finally, economists may also be able to contribute to the literature on theory choice within the situational analysis framework. No one knows the history of economics as well as economists, and the history of applied situational analysis is essentially the history of neoclassical economics.

NOTES

1. For the purposes of this discussion, Imre Lakatos's (1968, 1970) well-known distinction between "naive" and "sophisticated" falsificationism need not be considered.

2. For Popper such explanations must be based exclusively on the aims-desires-preferences of individuals, not groups, countries, classes, etc. In this regard, Popper is strictly a methodological individualist. Some students of Popper (e.g., Agassi, 1960, 1975) argue that institutions can be admitted into the discussion and still have the explanation remain a situational analysis explanation, as long as these institutions do not have aims. Others (e.g., Wisdom, 1970) argue that nonmethodological individualist yet situational explanations can be presented that use "emergent phenomenon." The complex question of the relationship between situational analysis and methodological individualism is beyond the scope of the current discussion. In what follows, the only situational explanation considered will refer to the actions of individuals. This is consistent with what many believe to be the "hidden agenda" of neoclassical economics (Boland, 1982).

3. Popper's statements on the rationality principle are scattered (1961, pp. 149-52; 1966, pp. 96-98; 1967, pp. 143-45; 1972, p. 179; 1976b, pp. 117-18). In addition to Koertge (1974), particularly clear statements are provided in Agassi (1960, p. 244; 1975, p. 146), Farr (1983, pp. 166-69), Koertge (1975, pp. 439-41; 1979a, pp. 86-87), Latsis (1976b, p. 21; 1983, p. 123), and Watkins (1970, p. 172).

4. Popper has consistently defended methodological monism. Popper (1961), for instance, can be viewed as simply one long argument in favor of this methodological position.

5. Popper has never claimed that "metaphysical" is identical to "immune to rational criticism." There may well be good "Popperian" reasons why one metaphysical theory is better than another. See Koertge (1978, pp. 272-75).

6. As quoted above, similar remarks were made in Popper's intellectual autobiography (1976b, pp. 117-18) and also (1976a, p. 102). This close relationship between economics and the rationality principle is viewed negatively by some of Popper's critics. For instance, Latsis states, "Popper's writings on the rationality principle mistakenly represent a philosophical reconstruction of one research program whose main domain has been economic theory as *the* method of the social sciences" (1983, p. 143).

7. Various perspectives on the role and status of the rationality (or maximization) postulate in economics are surveyed in Caldwell (1982, pp. 146-69). This discussion has recently been reopened by Boland (1981, 1983) and Caldwell (1983b).

8. Rationality Principle.

9. Lakatos (1968) and appendix to Lakatos (1970).

10. Again, Lakatos's distinctions among various forms of falsificationism are unnecessary here, although $Popper_{n0}$, $Popper_{n1}$, etc. would seem to follow naturally.

11. An infraction, by the way, for which the current author can also be indicted (see Hands, 1984a). It should be noted that Caldwell (1981), in his review of Blaug (1980a), suggested that Popper was not a strict falsificationist with respect to economics. He evidently chose not to elaborate on this suggestion in his later work.

12. This is Hutchison's translation of Popper (1967, pp. 145-46), with Hutchison's italics. Hutchison actually refers to "Popper (1967, pp. 145-48)," but the "8" must be a typographical error.

13. Stanley Wong's (1978) study in situational analysis has been neglected because it does not concern the general relationship between the method of situational analysis and the method of economic theory. Wong uses a form of situational analysis as a method of understanding and appraising the contribution of a particular economist (Paul Samuelson) rather than the general methodology of economics.

14. Austrian methodology has recently received a lot of attention (Blaug, 1980a, pp. 91-93; Greenfield and Salerno, 1983; Hutchison, 1981), although even Austrians disagree about what it is (see Kirzner, 1976; Rothbard, 1976). Even Robert Nozick's (1977) attempt to characterize the Austrian position has been severely criticized (Block, 1980). Thus far, much less has been written regarding the relationship between Hayek and Popper (Caldwell, 1983a; Gray, 1982; Hutchison, 1981). A detailed critical survey of the Austrian methodological literature is provided in Caldwell (1984b).

15. When combined with the assumption of diminishing marginal utility.

16. "If a given action or belief appears to be irrational, always blame your model on the agent's situation, not the Rationality Principle" (Koertge, 1975, p. 457). Also see Farr (1983, p. 170), Koertge (1974, p. 201; 1979a, p. 93), and Watkins (1970, p. 173).

17. While Popper$_n$ has changed his definition of ad hoc a number of times (see Koertge, 1979b; and Musgrave, 1974), this case violates even the earliest and simplest.

As regards auxiliary hypotheses we propose to lay down the rule that only those are acceptable whose introduction does not diminish the degree of falsifiability or testability of the system in question, but, on the contrary, increases it. . . . If the degree of falsifiability is increased, then introducing the hypothesis has actually strengthened the theory: the system now rules out more than it did previously; it prohibits more. (Popper, 1968, pp. 82-83)

18. See Koertge (1978) and Watkins (1958, 1975 and 1978).

19. See the last paragraph of the section "Situational Analysis and the Rationality Principle," above.

7

Ad Hocness in Economics and
the Popperian Tradition

This paper was prepared for a symposium on "The Popperian Legacy in Economics" held in Amsterdam in 1985. The symposium was organized to honor Joop Klant's work in the history and philosophy of economics— although the paper topics ranged widely over a number of areas in the Popperian tradition as well as alternatives to it. The conference papers and portions of the surrounding discussion were published in a volume edited by Neil de Marchi (1988a). De Marchi's introduction to this volume provides an excellent overview of the conference and the many issues that were raised. My contribution (Hands, 1988) focused on the issue of ad hoc theory adjustment, in Popperian philosophy and in economic theory. Much of the paper is simply a survey of the Popperian philosophical literature but I also provide an examination of the way the term ad hoc is used in economic theory. Since "novelty" was originally introduced by Popper to avoid one particular type of ad hoc theory adjustment, the discussion of ad hocness in this chapter relates to the previous discussion of novelty in Chapters 4 and 5.

This chapter discusses ad hoc theory adjustment in two separate fields: economics and Popperian philosophy. It will be shown that there are two fundamentally different concepts of ad hocness within the Popperian tradition—one associated with Karl Popper himself and one with Imre Lakatos—and that these two different concepts are mirrored in the way the term is used by economic methodologists and economic theorists, respectively. The first section below will discuss Popper's use of the term *ad hoc* and its fundamental importance to the Popperian program.

In the second section below, Lakatos's three different notions of ad hocness will be reduced to two (one the same as Popper's and one uniquely Lakatosian), and the importance of each of these to the methodology of scientific research programs will be examined. Although these first two sections are surveys of the existing literature, they are necessary because this particular aspect of the Popperian (and Lakatosian) philosophy of science has been badly neglected in the recent literature on economic methodology. The third section below discusses the use of "ad hoc" by economic methodologists and economic theorists, and compares these uses with those in the philosophical literature. The conclusion will examine the methodological implications of the preceding three sections. It will be argued that proper emphasis on the different notions of ad hocness, and on the different groups that emphasize each use, has significant implications for economic methodology. These implications are particularly important to Lakatosian economic methodology.

AD HOCNESS: POPPER

There is a long tradition in scientific philosophy as well as in scientific practice that considers a new theory to be less than satisfactory if it is *designed solely to deal with a previously anomalous observation*. Such theories do not go beyond their predecessors and they are considered to be ad hoc. Although the term *ad hoc* has no universally accepted definition within the philosophy of science, it is most often reserved for a particular type of face-saving/falsification-avoiding theoretical adjustment. For instance, Larry Laudan's popular definition states, "A theory is ad hoc if it is believed to figure essentially in the solution of all and only those empirical problems which were solved by, or refuting instances for, an earlier theory" (1977, p. 115).

For an example of the problem of ad hocness, consider the following (hypothetical) case from economic theory. Suppose a macroeconomic theory T_1 predicts that, for any country whose money supply changes by the amount m, the price level in the following period will change by the amount p. Suppose further that, after this result has been successfully confirmed for several countries, a contrary result is obtained for country A; that is, T_1 is refuted by the evidence from country A. How can T_1 be saved? How can T_1 be adjusted so that the potential refutation is avoided? One approach is simply to modify T_1 in an ad hoc way to avoid the refutation. For instance, we could modify T_1 so it now says that, "for all countries except country A, m implies p." This new theory—call it T_2—is confirmed by all the countries that confirmed T_1, and in addition, it is confirmed in country A, where T_1 failed. Thus, it could be argued that, since T_1 was refuted and T_2 is more confirmed (or, in Popper's terms, more "corroborated"), the move from T_1 to T_2 constituted "scientific progress."

It seems obvious, however, that this type of theory modification should not count as progress. T_1 was refuted, but T_2 is not *really* any better. The change from T_1 to T_2 was patently ad hoc; it was contrived solely to protect the theory from an observed anomaly, and served no other purpose. Not only does this amount to "cheap success" (Worrall, 1978, p. 49); it also makes "progress" trivially easy.[1] By this technique, a successful theory can always be constructed out of one with a dismal track record. The history of human speculation abounds with such (ultimately unsuccessful) face-saving moves.[2,3]

The ad hocness problem is even more important to those within the Popperian tradition than it is to those advocating other philosophies of science. Historically it was Popper's reaction to precisely this type of ad hoc defensive stratagem that originally motivated his interest in the demarcation problem.[4] The entire Popperian approach to science is characterized by *bold* conjecture and *severe* tests—precisely the opposite of such ad hoc/defensive behavior. For Popper "the aim of science is to get explanatory theories which are as little *ad hoc* as possible: a 'good' theory is not *ad hoc*, while a 'bad' theory is" (1965, p. 16).[5]

Now while ad hoc adjustments are generally undesirable, one should not be too reckless in eliminating the ways that a theory can be adjusted in response to contrary evidence. As Popper explained in his autobiography, he "also realized that we must not exclude all immunizations, not even all which introduced *ad hoc* auxiliary hypotheses" (1976b, p. 42). Popper cites the modification of Newton's theory in response to the observed motion of Uranus as an example of an ad hoc adjustment that eventually contributed to the development of the Newtonian program. What is needed is a methodological rule that bars ad hoc adjustments in a way that is strict, but not too strict. Popper's solution to this problem is the requirement of *independent testability*.

By "independent testability" Popper simply means that a new theory should make testable predictions independently of the predictions made by its predecessor.[6] Or to put it more in the Popperian vernacular, the new theory should have *excess empirical content;* in addition to inheriting and correcting the potential falsifiers of the early theory, the new theory must have additional falsifiers of its own.[7] A new theory that meets this requirement will be bolder and "better testable than the previous theory: the fact that it explains all the explicanda of the previous theory, and that, in addition, it gives rise to new tests, suffices to ensure this" (Popper, 1965, p. 242). An independently testable theory cannot be ad hoc in the way discussed above.[8]

Let us see how the independent testability requirement might be applied to the earlier economic example. Suppose that further examination of the anomalous country A reveals that the government of this country was running an extremely large deficit during the years when theory T_1 failed to predict inflation accurately. Suppose further that a reexamination of the countries where T_1 had been successful reveals that the governments of these countries had budgets close to balanced during the years in question. These observations might suggest a new and much bolder theory, T_3. T_3 would predict the initial result (same as T_1) for countries with budgets close to balanced, and a different result—the one obtained in country A—for countries with large deficits. Notice that T_3 predicts the same results as T_1 where T_1 was successful, corrects T_1 for the case of country A, and provides independently testable predictions. For example, T_3 tells us that, for any as yet unexamined country, we should find different results during years when the country was at war (and thus running a large deficit) than during years when the country was at peace. T_3 has excess empirical content; it leads to "new testable consequences, suggested by the new theory and never thought of before" (Popper, 1965, p. 243).

The last few words of this quotation reveal another aspect of the Popperian response to the ad hocness question: the importance of *novel facts*. In the paper considered to be his most important on the topic of scientific progress (1965, pp. 215-50),[9] *Popper identified independent testability exclusively with the prediction of novel facts.*

> We require that the new theory should be *independently testable*. That is to say, apart from explaining all the explicanda which the new theory was designed to explain, it must have new and testable consequences (preferably consequences of a new kind); it must lead to the prediction of phenomena which have not so far been observed. (Popper, 1965, p. 241)

Popper admits that requiring the new theory to predict "what had never been thought of before" (1965, p. 243) may "sound strange" (p. 247) because it means that whether or not evidence counts in favor of a theory depends critically on "whether the theory is temporally prior to the evidence" (p. 247).[10]

Now *if* Popper's novelty requirement for non-ad hocness is strictly interpreted, it has the "unsatisfactory consequence" (Watkins, 1978, p. 34) that "known" evidence does not contribute to the confirmation of a theory, that is, that *only* novel facts matter.[11] It has been pointed out by a number of authors that such a novelty requirement is sufficient to rule out ad hoc maneuvers, but it is not necessary to do so.[12] For instance, in our example, if the evidence regarding inflation and governmental deficits were known before T_3 was proposed, those facts would not be novel and would not count as independent evidence for the new theory. Such implications have led several authors to propose less restrictive definitions of novel.[13]

In summary, then, for Popper good science is that which makes bold empirical conjectures and exposes these conjectures to severe tests. The ad hoc adjustment of a theory for the sole purpose of protecting it from such a refutation is precisely the opposite of what should occur in good scientific practice. Popper proposed independent testability as a requirement that would prevent such ad hoc adjustments. Because independent testability was associated with the prediction of novel facts, the two concepts—novelty and non-ad hocness—came to be regarded as synonymous.

AD HOCNESS: LAKATOS

In his presentation of the "Methodology of Scientific Research Programmes" (hereafter, MSRP),[14] Lakatos defines a particular step in the development of a research program (a series of theories with shared hard-core metaphysical presuppositions and heuristic recommendations) as *theoretically progressive* if it "has some excess empirical content over its predecessor, that is, if it predicts some novel, hitherto unexpected fact" (1970, p. 118), and *empirically progressive* if some of these novel facts are actually confirmed. Lakatos defines a research program as *progressive* (nondegenerating) if it is theoretically progressive at each step in its development, and empirically progressive at least "intermittently" (1970, p. 134; 1968, p. 170). Notice two things about these definitions. First, even theoretical progress requires the prediction of novel facts.[15] And second, Lakatos uses the expressions "predicts novel facts," "has excess empirical content," and "is independently testable" as perfect substitutes for one another.[16]

Although Lakatos's notion of progress is tied to the same testability/ novelty requirement that Popper used to eliminate ad hoc theory modifications, Lakatos's actual use of the term differs significantly from Popper's. In fact, Lakatos uses the term *ad hoc* in at least three separate ways.[17] The first of these, which he calls "ad hoc$_1$," corresponds precisely to the way the term was used by Popper (and above). A theory is ad hoc$_1$ if "there is no *independent test* possible for it"

(Lakatos, 1970, p. 175, n. 2). Thus Popper's requirement for non-ad hocness is the same as Lakatos's requirement for nondegeneracy,[18] and the following equivalency holds among these concepts: non-Popperian ad hoc = non-ad hoc_1 = theoretically progressive.[19] Lakatos, like Popper, makes non-ad hocness an essential component of his notion of progress in science.[20]

Lakatos's second definition of ad hocness, "ad hoc_2," is only a slight modification of ad hoc_1. He calls a theory (or problem shift) ad hoc_2 if it has excess content but *none* of this "excess content got corroborated" (Lakatos, 1970, p. 175, n. 2). This definition creates the following equivalency: non-ad hoc_2 = empirically progressive.[21] Though requiring a theory to be non-ad hoc_2 is technically more stringent than requiring it to be non-ad hoc_1, the difference is only one of degree; ad hoc_2, like ad hoc_1, is fundamentally related to Popper's notion of ad hocness. Both requirements—theoretical and empirical progressiveness—represent ways to prevent the type of face-saving/content-decreasing/defensive behavior that Popper sought to exclude. This is not the case, though, for Lakatos's third notion of ad hocness.

Lakatos's third notion of ad hocness, "ad hoc_3," is related to the concepts of a research program's *positive heuristic*. According to MSRP, the hard core of the research program is the set of metaphysical (irrefutable) propositions that define the program and remain essentially unchanged throughout its evolution. The positive heuristic of the program is a set of rules that specify the research plan of the program and its relationship to the hard core.[22] The positive heuristic tells scientists working in the program what types of projects to work on, what empirical tests to perform, and how to interpret the results of those tests.[23]

Given this definition of the positive heuristic, ad hoc_3 is characterized as follows: "The theory is said to be ad hoc_3 if it is obtained from its predecessor through a modification of the auxiliary hypotheses which does not accord with the spirit of the heuristic of the programme" (Zahar, 1973, p. 101).[24] The absence of such ad hoc_3 stratagems provides a dividing line "between 'mature science,' consisting of research programmes, and 'immature science,' consisting of a mere patched up pattern of trial and error" (Lakatos, 1970, p. 175). It also defines an *alternative type of progress*: "non-ad hoc_3 = heuristically progressive" (Zahar, 1973, p. 101, n. 1). The motivation for defining (and avoiding) this third type of ad hocness is that scientific progress should entail increasingly more unified and cohesive theories; it should not be achieved (though it could be if only ad hoc_1 and ad hoc_2 adjustments are excluded) "with a patched up, arbitrary series of disconnected theories" (Lakatos, 1970, p. 175). Banning ad hoc_3 adjustments guarantees the continuity of science and preserves its "organic unity" (Zahar, 1973, p. 105). Thus, when scientific research *programs* are being considered, the previously discussed requirements for *progress* must be revised to include *heuristic progress*.

> Within a given programme a theory represents progress if it satisfies three conditions: relatively to its predecessors, it should entail novel predictions; some of these predictions ought eventually to be confirmed; finally, the theory should be structured in accordance with the heuristic. (Zahar, 1983, p. 170)

This third type of ad hocness is clearly different from the type that concerned Popper (and was discussed above).[25] It is apparent that even an empirically progressive (non-ad hoc_2) adjustment might still be ad hoc_3. For instance, it could be that T_3, proposed in our economic example above, is ad hoc_3 even though it is not guilty of the other forms of ad hocness. Unless we know the macro research program in which the theory is embedded, and can show that something in the hard core or positive heuristic of that program suggests that deficits should matter, T_3 could be accused of being ad hoc_3. Lakatos clearly considered program continuity to be extremely important and something that was not guaranteed by Popper's attempts to exclude ad hoc adjustments. For Lakatos, real progress occurs when a research program achieves theoretical/empirical progress "while sticking to its hard core" (1970, p. 187).

Thus there are two separate notions of ad hocness within the Popperian tradition. Ad hoc_{1-2}, or Popperian ad hocness, is the traditional notion of modifying a theory for the sole purpose of avoiding falsification. It can be prevented by requiring independent testability (or testing), that is, the prediction (or confirmation) of novel facts. The second notion, due to Lakatos, is ad hoc_3, and it pertains to the continuity of the program in which the theory is embedded. Ad hoc_3 adjustments can be avoided by requiring that any change be consistent with the core and positive heuristic of the program. Armed with these two concepts of ad hocness, let us now turn to the way the term is used in economics.

AD HOCNESS IN ECONOMICS

The purpose of this section is to compare economists' use of the term *ad hoc* with its use in the philosophical literature: ad hoc_{1-2} and ad hoc_3. In determining the way ad hocness is used in economics, two separate cases will be considered. The first is the way that *economic methodologists* use the term when appraising economic theories, and the second is the way that *economic theorists* (particularly contemporary theorists) use the term in theoretical discourse. In both cases, what is desired is a generalization regarding "the" representative or paradigmatic use of the term. Given the variety of ways the term is used, particularly among theorists,[26] obtaining such a generalization is quite an empirical problem in itself. All that will be done in the current context is to posit a representative use and offer a few quotations/references in its defense. Hopefully this posit will prove to be an "empirically progressive generalization"; that is, it will "anticipate" the way the term is used by (as yet) unexamined economic theorists.

There should be no controversy regarding the first group; the evidence clearly suggests that economic methodologists (and philosophers writing about economics) use the term in the traditional Popperian manner. For these authors, a theory is ad hoc if it is deliberately patched up to avoid falsification. For instance, Mark Blaug (1980a) criticizes both human capital theory (p. 238) and the new economics of the family (pp. 242-43) as ad hoc in this traditional sense (despite Gary Becker's insistence to the contrary), and Terence Hutchison accuses Marxists of making "*ad hoc* adjustments and qualifications" (1981, p. 18) to protect the Marxian theory of capitalist development from its predictive failures.[27,28] As yet another example, Alexander Rosenberg argues that most of the theoretical developments in the history of neoclassical rational choice theory amount of "largely *ad hoc* qualifications and restrictions that have preserved the theory against a series of failures to empirically substantiate it" (1980, p. 79).

Like methodologists and economic philosophers, economic theorists also use ad hocness in a defamatory way. It is not uncommon in the recent theoretical literature to find one economist condemning other economists or their theoretical approach as ad hoc. This tendency is particularly pronounced in the rational expectations (or "new classical macro") literature. In this controversial area it seems that "everyone calls the theory of the other ad hoc" (Klamer, 1983, p. 111). For example, in defending their use of the Lucas model, Thomas Sargent and Neil Wallace state, "The advantage of Lucas's model is that ad hockeries are given much less of a role."[29] It is quite common in surveys of this literature to hear that the assumption of rational expectations is compelling "in comparison with *ad hoc* alternatives" (Perry, 1984, p. 404) or that adaptive expectations—the principal alternative to rationally formed expectations—are simply "*ad hoc* rules" (Begg, 1982, p. 29). Keynesian alternatives to the new classical macroeconomics are said to rely on wage and price stickiness, which is "simply an *ad hoc* assumption" (Olson, 1984, p. 299), or they are based on "plausible—but you certainly could say ad hoc—disequilibrium dynamics" (Robert Solow in Klamer, 1983, p. 139). In a few cases the accusatory arrow is even reversed; for instance, Frank Hahn argues that it is the rational expectations theorists who "have chosen just that *ad hoc* model that delivers the goods" (1983, p. 60).

Although accusations of ad hocness seem to fly more freely in the rational expectations literature than in other theoretical areas, they are not exclusive to it. One of the other areas where it occurs is in traditional (Arrow-Debreu) general equilibrium theory. It has been known since the 1960s that, in order to guarantee the stability or determinant comparative statics for a standard Walrasian general equilibrium system, it is necessary to impose quite restrictive mathematical properties on aggregate excess-demand functions. Such restrictions are often called "*ad hoc* specialization of excess demand functions" (Fisher, 1983, p. 13) or "ad hoc assumptions, imposed directly on the excess demand system" (W. Hildenbrand's introduction to Debreu, 1983, p. 26).

How then are these economic theorists using the term *ad hoc?* Are they using it as methodologists do, as a claim that the accused theory has been deliberately modified solely to avoid an empirical refutation? The answer clearly seems to be no. Consider what immediately follows the Sargent and Wallace quotation cited above; it continues thus: "ad hockeries are given much less of a role and, consequently the neutrality proposition he obtains is seen to be a consequence of individual agents optimizing behavior."[30] Similar statements are made regarding the ad hocness of adaptive expectations. For instance, David Begg states, "I argued that *ad hoc* rules such as Adaptive Expectations have the disturbing implication that they allow individuals to make systematic forecasting errors period after period" (1982, p. 29), and this "suboptimal use of available information is hard to reconcile with the idea of optimization" (p. 26).[31] The previously mentioned Keynesian rigidities are not questioned for empirical reasons but because the "assumption [of rigid wages or prices] is needed to generate the main Keynesian results, yet it is not derived from or supported by any analysis grounded in the motivational assumptions economists have found to be general applicable" (Olson, 1984, p. 299). Even Solow's disequilibrium dynamics are potentially ad hoc because they are "certainly not the solution of some vast intertemporal optimization problem" (Klamer, 1983, p. 139). These statements do not indicate that nonrational expectations models are ad hoc because they lack empirical novelty or independent testability (though they might); rather, they are ad hoc because they are not derived from individual optimizing behavior. Nonrational expectations models are *accused of being ad hoc$_3$,* *not ad hoc$_{1-2}$;* according to their critics, the assumptions they require *do not follow from the core or positive heuristic of the neoclassical/ individual optimization research program*, and this is the source of their ad hocness.

What about criticism going the other way? That is, what about economic theorists who accuse the rational expectations models of being ad hoc? Are they concerned with ad hoc$_3$ as well? Certainly this is the case for Hahn. The basis of Hahn's ad hocness charge is that Lucas-type models assume equilibrium but fail to explain the presence of that equilibrium on the basis of rational optimizing behavior. According to Hahn, "much importance is attached to rationality until it comes to price changes: then anything goes. For the Lucasians, prices change to keep Walrasian markets cleared by a mechanism that is entirely secret in the Lucasian mind" (1983, p. 54).[32] The accusation here, as in the case of nonrational expectations models, is that the theory in question fails to follow from the positive heuristic of the neoclassical research program, not that it is ad hoc$_{1-2}$.[33]

What about the situation outside of the rational expectations literature? Is the concern over ad hocness only a rational expectations phenomenon, or is it a more general phenomenon in economic theory? Certainly if the general equilibrium case cited above is any indication, the concern is disciplinewide. Excess demand restrictions that guarantee stability and comparative statics (gross substitutes being the most common) are ad hoc because they do not follow from the standard assumptions about utility-maximizing agents; that is, they are not implied by the maximization of a strictly quasi-concave differentiable utility function subject to parametric prices and income. Since it is empirically obvious that not all goods actually exhibit properties like gross substitutability, it hardly seems possible that such assumptions were proposed to absorb a potential falsification. In addition, conditions like zero degree homogeneity and Walras's law are *not* considered ad hoc (though empirically questionable) simply because they *do* follow from the standard assumptions of consumer choice theory (and are therefore non-ad hoc$_3$ with respect to the neoclassical program).

Thus, for economic theorists, the sin of ad hocness seems to be infidelity to the metaphysical presuppositions of the neoclassical program, rather than face-saving adjustments in response to recalcitrant data. This is not to say that economic theorists do not actually adjust their theories in an ad hoc$_{1-2}$ manner, but only that recent theorists seem to consider ad hoc$_3$ness to be a more damning criticism. The methodological implications of this result will now be examined.

IMPLICATIONS AND CONCLUSION

One response to the realization that economic theorists and economic philosophers use the term *ad hoc* in different ways would be the so-called rhetorical response.[34] According to this interpretation, these differences merely reflect the fact that methodologists and theorists are engaged in different types of discourse; they attempt to persuade different audiences and use different means to do so. Philosophers and methodologists are trying to persuade an audience convinced of the cognitive superiority of "science" that certain areas within economics do not measure up to those exacting standards. Economic theorists, on the other hand, are trying to persuade an audience convinced of the methodological superiority of the neoclassical approach that some particular economic theories do not measure up to those standards. Based on such a rhetorical view, the interesting things about ad hocness pertain to why certain individuals use it in certain ways, what it is about the conversational context that requires one use over another, and what political/sociological/psychological factors influence the discourse of the two groups.

Although this rhetorical view seems fine as far as it goes, it does not exhaust the implications of the ad hocness issue. The differences among the various ways ad hocness is used have methodological implications that go beyond these rhetorical concerns. For one thing, the preceding discussion can help economic methodologists provide a better rationalization of theoretical developments, particularly from a Lakatosian perspective. A good example of this is provided in Maddock (1984), a Lakatosian reconstruction of the rational expectations literature. At one point, Maddock notes an apparent "methodological inconsistency" in the work of Sargent and Lucas: "they rejected ad hoc adjustments to their models which could overturn the basic propositions, but made their own ad hoc adjustments in their endeavor to corroborate statistically those same propositions" (1984, p. 301). This methodological inconsistency seems to dissolve when it is realized that different types of ad hocness are involved in the two cases of adjustment. In the first case, the "ad hoc adjustments" rejected by rational expectations theorists are ad hoc$_3$ adjustments that would introduce parameters not motivated by individual optimization. In the second case, where the rational expectations theorists "make their own ad hoc adjustments," the ad hocness is of the more traditional ad hoc$_{1-2}$ type. In this second case, as Maddock clearly documents, there was a rather blatant attempt to account for the empirical fact of "persistence" by adding a (thoroughly ad $hoc_{1\text{-}2}$) lagged unemployment (or income) term to the right-hand side of the Lucas supply function.

The two uses of ad hocness and the relative importance that theorists attach to each can also help explain Lucas's response to this apparent inconsistency. Rather than provide a new theory that would fit the data without being ad $hoc_{1\text{-}2}$, Lucas's response was to provide a formal general-equilibrium model in which persistence could be derived from informational imperfections. As Maddock explains,

> the introduction of a lagged income term into the aggregate-supply equations which had appeared ad hoc could now be justified as arising from within a consistent general equilibrium model with information differentials. As this paper stood it made no new empirical prediction, but it did lessen the impact of a criticism. (Maddock, 1984, p. 302)

What it did, of course, was to lessen the criticism from other theorists that the supply function was ad hoc_3. Maddock claims that from a Lakatosian perspective Lucas's move was "defensive" (1984, p. 302); but maybe not. The theory certainly did not predict any novel facts, but maybe it was still "going forward" in a Lakatosian sense. What we have is a case where the lagged income term was both ad $hoc_{1\text{-}2}$ and ad hoc_3. According to Lakatos's MSRP, *both types* of ad hocness *must be eliminated* to avoid degeneracy.[35] Nothing in Lakatos's work specifies which type of ad hocness should be eliminated first.

This brings up a second but related point about the use of ad hoc_3ness in economics. Although heuristic progress (non-ad hoc_3ness) seems to be what most concerns economic theorists, it is almost never mentioned in the extensive literature that applies the MSRP to economics.[36] If heuristic progress were seriously considered, it might substantially change the nature of Lakatosian reconstructions in economics. On the pro-Lakatosian side, as shown in the Lucas case above, it may be possible to reconstruct certain theoretical developments as rational (or nondegenerating) by Lakatosian standards, where they would not be if only ad $hoc_{1\text{-}2}$ were considered. Lakatos *does* require heuristic progress, and it plays an important role in Lakatosian reconstructions of physical science.[37] If the profession's well-known infatuation with maximization can be reinterpreted as heuristic progress within a neoclassical research program that has individual optimization as a hardcore proposition, the entire history of economics may appear more progressive in Lakatosian terms. Also on this same side, it appears that economic theorists themselves have something like a general neoclassical research program clearly in mind. They seem to have already formed a hard core and a positive heuristic, and to know when the latter is violated. This supports those economists attempting to reconstruct most of neoclassical economics as one big research program.[38] On the anti-Lakatosian side, it seems that theorists' revealed preference for heuristic over theoretical/empirical progress (as Lakatos defines these terms) will make it hard to reconcile the history of economic thought with Lakatos's requirement that novel facts be predicted (though they need not be confirmed) at each stage in the program's development. This preference for heuristic progress may explain the paucity of novel facts in the history of economic thought.[39] It may be necessary to modify the MSRP to suit the needs of economics by "weighting" these two types of progress and finding an optimal way of making trade-offs between them. If this is to be done in a way consistent with the actual practice of economic theorists, it seems that a relatively large weight should be given to heuristic progress. In any case, it appears that philosophers and economists taking a Lakatosian approach to economics (as well as those criticizing it) should seriously reconsider the question of heuristic progress (non-ad hoc_3ness) and its role in the evaluation of economic theory.[40]

Finally, in ending this investigation of implications of ad hocness, it is important to recall our discussion of Popper from an earlier section. Popper was concerned with ad hoc_{1-2}, with the deliberate modification of a theory for the sole purpose of avoiding a refutation. The concern with independent testability and novel facts only came about as a possible way of avoiding ad hoc_{1-2} theory adjustment. In other words, novel facts are *not* fundamentally interesting to Popper. They are only derivative; they are interesting because they help solve the problem of ad hoc_{1-2} theory adjustment—a problem that *is* fundamentally interesting. Thus, even if we accept Popper's concern over ad hoc_{12} adjustments, novelty need not be as significant as recent work would indicate.[41] The elevation of novelty to an independently interesting concept is due to Lakatos. Although Lakatos's interest in novel facts initiated from the same basic concern as Popper's, it seems to have turned into a novel fact fetishism that lost sight of the original problem. It may be time in economic methodology to reexamine Popper's problem: the issue of ad hoc theory adjustment in the traditional sense, and what might prevent it. Do traditionally ad hoc adjustments often occur in economics? And if so, how are they problematic? Are there rules that would prevent such adjustments as they occur in economics? Such questions would bring us back to the traditional issue of ad hocness and Popper's problem—something that seems to have been lost in the shuffle of recent methodological discussion.

NOTES

1. For this reason, John Watkins (1984) refers to the non-ad hocness requirement as an "antitrivialisation principle." The principle states that "any philosophical account of scientific progress must be inadequate if it has the (no doubt unintended) implication that it is always *trivally easy* to make theoretical progress in science" (Watkins, 1984, p. 166).

2. Grunbaum (1976b, pp. 329-30) cites historical examples.

3. It should be noted that, although the undesirability of ad hoc theories is being presented as obvious, not all philosophers of science would agree. Larry Laudan (1977, pp. 114-18), for instance, finds no difficulty with this type of ad hoc modification, claiming that to condemn such adjustments puts "an epistemic premium on theories which work the first time around" (p. 117). Adolf Grunbaum (1976b, pp. 358-61) takes a similar position.

4. Popper (1965, pp. 33-37; 1976b, pp. 41-44).

5. The "ad hoc" in italics here is original, as it is within quotations throughout this chapter.

6. Popper (1965, pp. 217-20, 240-48; 1972, pp. 191-205).

7. Recall that Popper defines the "empirical content" of a statement as the set of all its potential falsifiers (see Popper, 1965, pp. 332, 385; 1968, pp. 113, 119-211; 1976b, p. 26; and Lakatos, 1970, p. 111). Although Popper's definition of empirical content is sufficient for our present purposes, it should be noted that many, including some neo-Popperians, find it to be problematic. Watkins (1984, pp. 167-83) discusses the difficulties of the Popperian definition (actually, definitions) and offers his own modification called the "comparative testability" criterion.

8. Although independent testability is sufficient to rule out cases like T_2 in the preceding economic example (above in the text), it is not sufficient to rule out every type of ad hoc adjustment. To really guarantee that the new theory is non-ad hoc, Popper also requires that some of the excess content actually *be confirmed*.

> This becomes clear if we consider that it is always possible, by a trivial stratagem, to make an *ad hoc* theory independently testable, *if we do not also require that it should pass the independent tests in question:* we merely have to connect it (conjunctively) in some way or other with any testable but not yet tested fantastic ad hoc prediction which may occur to us (or to some science fiction writer). (Popper, 1965, p. 244)

9. And to a lesser extent in "The Aim of Science" (Popper, 1972, pp. 191-205).

10. Not everyone writing on Popper's transition from non-ad hocness to novel facts documents the transition in exactly the same way (see Musgrave, 1974, pp. 3-12; Watkins, 1978, pp. 33-36; 1984, pp. 288-300; and Worrall, 1978, pp. 45-51). Lakatos seems to argue (1978a, p. 172) that there was no

transition at all, that Popper always subscribed to the view that non-ad hocness is equivalent to the prediction of novel facts.

11. Although this is not the only way that Popper's discussion of novelty can be interpreted, it is a quite common interpretation. For instance, John Worrall says, "The Popperian account of empirical support says that a theory is supported by any fact which it describes correctly and which was first discovered as a result of testing this theory; and that a fact which was already known before the theory's proposal does not support it" (1978, p. 46).

12. Musgrave (1974, p. 12) and Worrall (1978, p. 49).

13. This changing view of novelty is discussed briefly in Hands (1985a, pp. 6-7). Alternative definitions have been provided in Gardner (1982), Musgrave (1974), Watkins (1984), Worrall (1978), and Zahar (1973). Lakatos's emphasis on novelty (discussed in the next section) has been an important factor in the development of this literature.

14. Lakatos (1968, 1970).

15. Based on what Lakatos actually wrote, this is not literally correct. He says that a change is theoretically progressive *if* it predicts novel facts; he does *not* say *only if* it predicts novel facts. Thus the door is (formally) left open for other things that would be sufficient for theoretical progress. Although this is an interpretation that can be defended on the basis of what Lakatos actually wrote, it is certainly not the standard interpretation. The standard interpretation is (as stated in the text) that progress *requires* novel facts; that is, "if and only if" is read in, rather than "if." The standard interpretation is probably more consistent with the overall spirit of the MSRP.

16. For example, he says "produces novel facts (that is, it is 'independently testable')" (Lakatos, 1970, p. 126) and "theories which had no excess content over their predecessors (or competitors), that is, which did not predict any *novel* facts" (p. 175, n. 2).

17. This follows his presentation in Lakatos (1978a), a paper originally published in 1968.

18. Actually, of course, Popper requires some confirmation of this excess content (see note 8 above) and Lakatos requires at least intermittent confirmation.

19. This equivalence is implicit in Lakatos (1968, 1970); it becomes explicit in later "Lakatosian" works such as Zahar (1973).

20. "The problem of when a hypothesis-replacement is 'ad hoc,' i.e., irrational, degenerating, bad, has never been discussed with more attention and detail than by Popper and myself" (Lakatos, 1978b, pp. 221-22).

21. Actually, this identity requires a slight modification of Lakatos's definition; see Zahar (1973, p. 101, n. 1).

22. Lakatos (1968, pp. 170-73; 1970, pp. 134-38) and Worrall (1978, p. 59). Worrall (1978, p. 69, n. 35) provides a partial list of the things that the positive heuristic might include.

23. As an economic example, in E. Roy Weintraub's specification (1985a) of the neo-Walrasian research program, the positive heuristic consists of propositions like "go forth and construct theories in which economic agents

optimize" and "construct theories that make predictions about changes in equilibrium states."

24. Elie Zahar is quoted here simply because his statement is more straightforward than Lakatos's, which must be extracted from his discussion on pp. 174-77 and 182-87 of his 1970 paper.

25. Though Popper did require a new theory to be "deeper," to have "a certain coherence or compactness" (1972, p. 197) or "simplicity" (1965, p. 241). Although these concepts are related to Lakatos's non-ad hoc_3, Popper was pessimistic about formalizing such requirements and never provided more than an intuitive discussion. Nor did Popper ever directly relate this coherence and simplicity to his notion of ad hocness. Recently, though, such concepts have been analyzed more rigorously (and related to ad hocness) by neo-Popperians—especially Watkins (1984).

26. Not only economic theorists but natural scientists as well "use the term *ad hoc* to cover a multitude of sins. A theory may be called *ad hoc* because it is unaesthetic and clumsy, because it is arbitrary and uninteresting, or because it is wildly implausible" (Koertge, 1978, p. 267). Grunbaum (1976b, p. 361) cites a quite extensive list of scientific uses of ad hocness (compiled by Gerald Holton).

27. Similar criticisms of Marxian economics are made by Blaug (1980b).

28. Although the ad hocness that concerns economic methodologists always seems to be of the Popperian type, it is interesting that Popperian methodologists such as Blaug, Hutchison, and Klant explicitly discuss the disease itself—that is, ad hocness—whereas Lakatosian methodologists are more likely to focus exclusively on the cure, that is, "excess content" and "novel facts." In either case, there is seldom any recognition that the issue is fundamentally the same.

29. From Sargent and Wallace (1976), quoted by Maddock (1984, p. 295).

30. Ibid.

31. In Begg's introduction to his nonrational expectations model he states, "We shall work with an ad-hoc specification of a macroeconomic model, rather than attempt to derive such a model from explicit microeconomic foundation" (1980, p. 294).

32. Making a similar point, Hahn also writes,

I am in Lucas's methodological spirit when, instead, I propose that prices are flexible when there are no obstacles to price change when it is to someone's advantage to do so. More formally, prices in a given theory are flexible when their formation is endogenous to the theory.... Now as a matter of fact, prices in the Lucasian world are not properly endogenous to the fundamental theory, because there is no theory of the actions of agents that explains how prices come to be such as to clear Walrasian markets. (Hahn, 1983, p. 49)

33. Following a time-honored Lakatosian tradition (Lakatos, 1971a, p. 107), theorists who have "misbehaved" relative to the preceding reconstruction are consigned to footnotes. One such case is Tobin (1980). In his discussion of

Lucas's model, James Tobin accuses (p. 42) it of being ad hoc in the traditional (ad hoc$_{1-2}$) sense. In response, Lucas fully admits that it is. "If ever there was a model rigged frankly and unapologetically to fit a limited set of facts, it is this one" (Lucas, 1981, p. 563).

The reader is reminded that the argument in the text tries to offer only an empirically progressive generalization about the way economic theorists use the term *ad hoc*, not a claim that it is used in no other way.

34. Arjo Klamer (1984) provides such a rhetorical view. For him the rational expectations theorists' use of ad hoc$_3$ is an "epistemological argument" that serves "the new classical economists well to ward off the criticism that their assumptions are unrealistic" (Kramer, 1984, p. 283).

35. "I define a research programme as degenerating even if it anticipates novel facts but does so in a patched-up development rather than by a coherent, pre-planned positive heuristic" (Lakatos, 1971a, p. 125, n. 36).

36. A partial list is provided in Hands (1985a), but many more works have appeared since that paper was written (e.g., Maddock 1984, and Weintraub, 1985a).

37. For instance, Zahar (1973) argues that Einstein's program superseded Lorentz's program because it was heuristically more progressive.

38. For example, Weintraub (1985a).

39. Although the absence of novel facts is a theme in Hands (1985a), in all fairness it should be noted that "paucity" does not imply "nonexistence." Maddock (1984) and Weintraub (1988a), for instance, both find novel facts in the economic programs they discuss. It may be just an interesting coincidence, but whenever philosophers in the Popperian tradition discuss authors who are critical of novel facts—that is, those who think that facts are facts, regardless of whether they are "known" or "unknown"—the two "philosophers" most often cited are John Stuart Mill and John Maynard Keynes (Lakatos, 1970, pp. 123-24; 1978a, p. 183; Musgrave, 1974, p. 2; Popper, 1965, p. 247).

40. It is interesting to note that in Lakatos's few offhand comments about social science (e.g., 1970, p. 176, n. 1), he considers its immaturity—that is, its lack of heuristic progress—to be its primary deficiency. It may be that economists' seemingly irrational concern with "sticking with the program" (even at the expense of novel facts) is precisely what makes economics the most mature of the social sciences.

41. Recall (from note 4) that not all philosophers think that ad hoc$_{1-2}$ness is a real problem. It can also be added that, within the Popperian tradition, not everyone who thinks that ad hoc$_{1-2}$ness is a problem believes that predicting novel facts is the way to solve it (Koertge, 1978, p. 269).

8

Falsification, Situational Analysis, and Scientific Research Programs: The Popperian Tradition in Economic Methodology

This chapter was originally published in a methodology volume edited by Neil de Marchi (1992). The purpose of the volume was to provide substantive surveys of a number of different approaches to economic methodology. My assignment was the Popperian tradition. The paper critically examines all three aspects of the Popperian approach to economic methodology: falsificationism, situational analysis, and Lakatos's MSRP. While it is principally a survey—a general overview of the "state of the literature" on the Popperian tradition—the paper is also an interpretative essay. This is particularly the case in the discussion of situational analysis, which goes substantially beyond the presentation in Chapter 6. The original paper appeared with a comment by Mark Blaug (1992) and my response (Hands, 1992b).

Perhaps no other philosopher or philosophical school has influenced economic methodology as much as Karl Popper. From the economics profession's introduction to falsificationist ideas in Hutchison (1938), to the recent spate of Lakatosian case studies in the history of economic thought,[1] few major issues in economic methodology have been discussed without a substantial Popperian voice.[2]

It is the purpose of this essay to reexamine critically this Popperian influence in economic methodology. The presentation will be in three sections, each corresponding to one of the three main points of contact between the Popperian tradition and the literature on economic methodology. The first section examines *falsificationism*; Popper's well-known approach to the philosophy of natural science. The second section discusses *situational analysis*, Popper's less well-known approach to the social sciences. The final topic considered is the work of Imre Lakatos and how it has been applied to the history of economic thought. While Lakatos's philosophical position differs substantially from Popper's, his work fits comfortably enough into the general Popperian tradition included in this reexamination of Popper. In each of the three areas, the focus will be on the literature that explicitly concerns economics; there will not be any effort to discuss general philosophical arguments or evaluations based on other scientific disciplines. Throughout the paper, survey material is provided to help familiarize the reader with the relevant literature, but the paper is not an exhaustive survey of any of the three individual topics.

FALSIFICATIONISM

No doubt, economists, philosophers, and members of the academic community more generally, know Karl Popper best for his falsificationist approach to the philosophy of science. First presented in *Logik der Forschung* in 1934 (English translation, Popper, 1968), falsificationism represents Popper's approach to the growth of knowledge as well as his solution to (or dissolution of) the traditional problem of induction. It is for his falsificationism that Popper claims responsibility for the death of logical positivism.[3]

Actually, Popperian falsificationism is composed of two separate theses: one on demarcation (concerned with demarcating science from nonscience), and one on methodology (concerned with how science should be practiced). The *demarcation* thesis says that for a theory to be "scientific" it must be at least *potentially falsifiable;* that is, there must exist at least one empirical basic statement that is in conflict with the theory.[4] This potential falsifiability is a logical relationship between the theory and a basic statement. In particular, the demarcation criterion does not require that anyone has actually tried to falsify the theory, but only that it would be logically possible to do so. Over the years Popper's demarcation criterion has been the subject of an extensive debate in the philosophical literature; however, demarcation does not seem to be the main issue in economic methodology.[5] For economists who advocate a falsificationist position, the most important issue is methodology, not demarcation; and Popperian *methodology* requires the *practical (rather than merely the logical) falsifiability* of scientific theories.

Briefly, and neglecting a number of philosophical issues, Popper's falsificationist methodology requires scientists to search for scientific knowledge in the following way. First, start with a scientific problem situation—something requiring a scientific explanation. Second, propose a bold conjecture that might offer a solution to the problem. Third, severely test the conjecture by comparing its least likely consequences with the relevant empirical data. The notion here is that a test is more severe the more prima facie unlikely the consequence tested; the theory should be forced to "stick its neck out," to "offer the enemy, namely nature, the most exposed and extended surface."[6] Fourth and finally, the last move in the game depends on how the theory performed during the third testing stage. If the implications of the theory were not supported by the evidence, then the conjecture is falsified and it should be replaced by a new theory that is not ad hoc relative to the original.[7] If the theory was not falsified, then it is considered corroborated by the test and it is provisionally accepted. It should be noted that, given Popper's fallibilism, this acceptance remains provisional forever; the method does not guarantee that the surviving theory is true, but only that it has faced a tough empirical opponent and won.

There seem to be a number of reasons why such a falsificationist approach to science might appeal to economic methodologists. If the task of economic methodology is viewed (as it has been until quite recently) as "choosing" among various philosophies of natural science in order to "apply" one to economics, then Popperian falsificationism has some clear advantages over anything that might be borrowed from the positivist tradition. For one thing, falsificationism is eminently more straightforward and intuitive than the inductive logic of the later logical empiricists. Perhaps more important is the fact that Popper's falsificationism is truly a methodology. Unlike philosophers in the positivist tradition, Popper's main focus was not to provide an epistemic justification for the knowledge claims of science. Popper's falsificationist goal was the more mundane task of characterizing a set of rules (a method) that would allow us to learn from experience. This distinction between methodology and justification seems to be critical to an understanding of the influence of Popperian philosophy on the economics profession. The reason is that by and large the positivist tradition was *not* a tradition of methodological rule making. Most logical empiricist philosophers were firmly convinced that science proceeds by induction; their philosophical task was to *justify* that procedure. Such a justificationist philosophy of science provides little or no guidance to an economics profession in search of scientific rules. The traditional question for economic methodology is not to provide a philosophical justification for scientific practice, but rather to find a set of rules that can be followed so economists can do whatever it is that scientists do.[8]

Another reason for the support of Popperian falsificationism among economists is that it seems to solve (or dissolve) the old induction versus deduction debate in economics; it provides a tidy way out of the *methodenstreit* without making either side the overall winner. Consistent with the apriorist-deductivist tradition in economics, the falsificationist method would allow hypotheses based on introspection and/or the presupposition of rational action. Consistent with the historical-inductivist tradition in economics, falsificationism requires empirical testing and discipline by the data. Popperian falsificationism seems to allow the profession to take advantage of what is best in each of these traditional approaches to economic research. It is permissible to leap to conjectures about economic behavior without the extensive accumulation of empirical observations that would be required if only inductive generalizations were allowed, while at the same time (unlike the Misesian approach) the facts do matter and acceptable hypotheses must survive severe tests. This "best of both worlds" property makes falsificationism a natural philosophical companion to the Marshallian tradition in economics—a characteristic that probably contributes to its support by many economic methodologists.

In addition to these philosophical issues, there are possibly some forces of attraction that should be classified as "sociological" (and/or personal, and/or ideological). In particular, Popper's direct influence on a number of influential LSE economists[9] and his long-standing relationship with F. A. Hayek may have contributed to Popper's popularity among economists. Certainly, citations originating from these two sources made Popper's name familiar to many economists who would not have otherwise been aware of his work. Finally, it is possible to find an ideological connection. Popper's own work in social and political philosophy[10] is decidedly antihistoricist and anti-Marxist—views that are (at the very least) not inconsistent with those of most mainstream economists. These sociological and/or ideological factors do not directly support a falsificationist methodology for economics. Rather, they explain why economists might consider Popper an "acceptable" philosopher and thereby (since falsificationism is Popper's most well-known view) lend indirect support to the falsificationist position in economics.[11]

Now despite all of these reasons why falsificationism might be a desirable methodology for economics, the fact is that *falsificationism is seldom if ever practiced in economics*. This seems to be the one point on which recent methodological commentators generally agree.

In fact, this (empirical) claim is supported at length by the case studies in Blaug (1980a), a work that consistently advocates falsificationism as a normative doctrine. The disagreement between critics and defenders of falsificationism is *not whether it has been practiced*—basically it has not—but rather *whether it should be practiced*. The real questions are whether the profession should try harder to practice falsificationism though it has failed to do so in the past, and whether the discipline of economics would be substantially improved by such falsificationist practice.[12]

One way to answer such queries regarding the appropriateness of falsificationism in economics is to consider the appropriateness of Popper's falsificationist methodology as a *general* approach to the growth of scientific knowledge. Falsificationism may not be appropriate for economics even if it is a good model for the growth of knowledge in natural science, but if it fails in natural science then its usefulness in economics is surely in doubt. Unfortunately, such an excursion into the vast philosophical literature criticizing Popperian falsificationism is beyond the scope of the current essay.[13]

Rather than delving into the more general philosophical literature, I will simply list some of the criticisms of falsificationism that can and have been raised explicitly within the context of economics. These criticisms may overlap with more general philosophical concerns; but even so, only economics will be explicitly discussed here. The list is not exhaustive, but it does capture many of the problems exhibited by a falsificationist methodology in economics. They are not listed in any particular order of importance.[14]

- The Duhemian problem[15] and related issues pose insuperable problems for falsificationist practice in economics. There are a number of reasons why this is the case.

    First, the complexity of human behavior requires the use of numerous initial conditions and strong simplifying assumptions. Some of these restrictions may actually be false, such as the differentiability of production functions or the infinite divisibility of commodities. Some of these restrictions may be logically unfalsifiable, such as the assumptions of eventually diminishing returns or eventually decreasing returns to scale. Still others of these assumptions may be logically falsifiable but practically unfalsifiable, such as the completeness assumption in consumer choice theory. And finally, most of these restrictions are extremely difficult to test for because of the absence of a suitably controlled laboratory environment. The presence of such a variety of restrictions makes it virtually impossible to "aim the arrow of *modus tollens*" at one particular problematic element of the set of auxiliary hypotheses when contrary evidence is found.

Second, in addition to these problems with auxiliary assumptions, there is not a clear consensus regarding the empirical basis in economics. It is always possible to argue that what was observed was "not really" involuntary unemployment or "not really" economic profit, and so forth. Unlike positivism, Popper's philosophy does not require the empirical basis to be incorrigible, but he does require the empirical basis to be a generally accepted convention.[16] In economics, even such a conventionally accepted empirical basis is frequently absent.

Third, it should be noted that social sciences can have feedback effects of a type that do not exist in the physical sciences. The test of an economic theory may itself alter the initial conditions for the test. Conducting a test of the relationship between the money supply and the price level may alter expectations in such a way that the initial conditions (which were true "initially") are not true after the test (or if the "same" test were conducted again).[17]

- The qualitative comparative-statics technique used in economics makes severe testing very difficult and cheap corroborational success too easy. Even with the auxiliary assumptions discussed in above, it is still frequently the case that the strongest available prediction is a qualitative comparative-statics result, which only specifies that the variable in question increases, decreases, or remains the same. Since predicting the correct sign of change in a parameter is always much easier to accomplish than predicting the correct sign and magnitude, this qualitative comparative-statics technique generates theories that are low in empirical content, have few potential falsifers, and are difficult if not impossible to test severely. The result is often economic theories that are corroborated but trivial.[18]

- Popper's "admitted failure" (1983, p. xxxv) to develop an adequate theory of verisimilitude[19] also presents problems for a falsificationist methodology in economics. The problem of verisimilitude developed as an attempt to reconcile Popper's falsificationist methodology with his scientific realism. For a realist, the aim of science is to find "true" theories; according to falsificationism, scientific theories should be chosen if they have been corroborated by passing severe empirical tests. If the falsificationist method is to fulfill the realist aim of science, it should be demonstrated that more corroborated theories are closer to the truth; such a demonstration was the goal of Popper's theory of verisimilitude.

Actually a satisfactory theory of verisimilitude would serve Popperian philosophy in at least two separate ways. The first, mentioned above, would be to provide an epistemic justification for playing the game of science by falsificationist rules. This issue is extremely important for Popperian philosophy since it means that, without a theory of verisimilitude, there are philosophically "no good reasons" (Popper, 1972, p. 22) for choosing theories as Popper recommends. The second function of a theory of verisimilitude is more practical: it would provide some rules for choosing the "best" theory in troublesome cases. This is because a theory of verisimilitude would provide rules for discovering which of two theories has more verisimilitude, which is a better approximation to the truth. Thus if we had two theories and both had been falsified, we could choose the one with more verisimilitude. Notice that falsificationism without a theory of verisimilitude is of no help in such cases; since both are false, both are *out*. Similarly for cases involving a choice between a falsified but bold theory and a corroborated but modest theory.[20] Again, having a way to determine which is closer to the truth might allow us to choose a theory more consistent with the aims of science than simple falsificationist rules.

This second, more practical, function of the theory of verisimilitude can be extremely important for economic methodology. For all the reasons discussed above, and perhaps for others as well, economists are almost always faced with choosing between two falsified theories or between a bold falsified theory and a more modest corroborated one. If Popper's theory of verisimilitude had been a success and it could be added to the norms of simple falsificationism (both to justify the norms and to help in making the practical decisions of theory choice), then falsificationism might have an important role to play in economic theory choice. Without the link between severe testing and truthlikeness, the method seems to be of limited value in pursuing the realist aim of science.

• Popper's rules for progressive theory development (non-ad hocness) are often inappropriate in economics. Popper argues that, if a theory is to constitute "progress" over a predecessor, the new theory must be "independently testable"; it must have "excess empirical content," predict "novel facts."[21] This issue will be examined in more detail in the section on Lakatos's methodology below; but for now let it be said that, while progress of this Popperian type may sometimes be of interest to economists, often progress in economics is (and should be) much different. Often economists are concerned with finding new explanations for well-known stylized (nonnovel) facts; or alternatively, economists are concerned with explaining the same phenomena with fewer theoretical restrictions. What exactly constitutes "progress" in economic theory (or what should constitute progress) is a complex and ongoing question, but it is apparent that any suitable answer will require a much more liberal set of standards than those offered by strict Popperian falsificationism.

These criticisms do not bode well for a falsificationist economic methodology. Despite all of the reasons why a falsificationist methodology might be attractive to economists, it fails to provide an adequate set of rules for doing economics. Strict adherence to falsificationist norms would virtually *destroy all existing economic theory* and leave economists with a rule book for a game unlike anything the profession has played in the past. This high cost would be paid without any guarantee that obeying the new rules would result in theories any closer to the truth about economic behavior than the economic theories that are currently available.

However, denying that falsificationism provides the proper methodological rules for conducting economics does *not* mean that "the facts" should not matter in economic theory choice or that empirical testing is not important. This type of argument is quite a common red herring in the methodological discussion about falsificationism in economics. Popperian falsificationism is not, as some economists seem to think, generic empiricism; it is a *very* specific set of rules about how scientific inquiry should be conducted. Abandoning Popperian falsificationism as a methodology does not mean abandoning learning from experience.

SITUATIONAL ANALYSIS

While economic methodologists have long been concerned with Popperian falsificationism, Popper's views on situational analysis have only recently become an explicit part of the literature on economic method. The most likely reason for this neglect is the relative (to falsificationism) inaccessibility of Popper's own work on the topic. The staunchest supporters of situational analysis have been social science-oriented philosophers such as I. C. Jarvie who became familiar with the argument through Popper's lectures;[22] and while the central thesis was presented in Popper's work on social and political philosophy (1961, 1966), the clearest presentation of the argument is Popper (1967), a paper that has only recently been translated from the original French (Popper, 1985). Other presentations of the topic are scattered about in works such as Popper (1976a), a paper written as part of a debate with the Frankfurt school of sociology.

Situational analysis is Popper's methodological approach to the social sciences.[23] In fact he argues that situational analysis is the *only* method appropriate for the social sciences.[24] Now since economics is surely a *social* science, it is paradoxical that economic methodologists have focused almost exclusively on falsificationism—Popper's philosophy of natural science—and neglected situational analysis. This paradox, though explicable in terms of the relative inaccessibility of Popper's writings on situational analysis, is even more pronounced since situational analysis *is the method of economic analysis.*[25]

According to Popper's situational analysis, explanations of human behavior should proceed as follows. Suppose the problem is to explain why agent A engaged in some particular type of behavior, say X. The first step in explaining this behavior is to describe the "situation" of the agent at the time the behavior in question took place. This description of the agent's situation will normally include both subjective components (the agent's goals, beliefs, desires, etc.) as well as objective components (the physical and social constraints faced by the agent). The second step in the explanation is to provide an analysis of the situation, to specify what type of behavior would be appropriate (i.e., rational) given the agent's situation. The third step in the explanation is to add—and this is the key—the *rationality principle* (RP), which asserts that *all* individuals *actually act in a way that is appropriate* to their situation (i.e., they act rationally). This RP allows us to deduce the act of the agent from the description of his or her situation and from our analysis of what constitutes appropriate behavior. The RP is a bridge principle that connects the "situation" with an "action"; it "stands in for the 'law' which 'animates' the otherwise inert collection of situational features" (Latsis, 1983, p. 133).

Schematically, then, the situational analysis explaining why agent A did X has the following form.

I.	Description of the Situation:	Agent A was in situation S.
II.	Analysis of the Situation:	In situation S the appropriate (rational) thing to do is X.
III.	The RP:	Agents always act appropriately (rationally) given their situation.
IV.	Explanandum:	Therefore, A did X.[26]

It is easy to see that situational analysis is the standard method of microeconomics (and possibly macroeconomics based on neoclassical microfoundations). Economists first specify the situation of the agent (individual or firm) in terms of the preferences and/or technology and the relevant constraints (prices, income, factor constraints, etc.). Included in the description of the situation is some motivating consideration (maximizing utility, maximizing profit, etc.). They next deduce the appropriate behavior of the agent given the situation specified (buy more, buy less, increase production, decrease production, etc.). This second step constitutes most of what is called "economic theory," the formal deduction (often highly mathematical) of the appropriate behavior of a particular agent in a particular situation. Finally, if the economist's task is to explain an observed action, the RP is activated to connect the analysis of the situation with the action to be explained. If the task is "pure theory," then this latter step is neglected, and the "theoretical result" involves technically deducing step II from a hypothetical situation in step I. Comparative statics results are obtained by simply performing the deduction from steps I to II twice, with a slight change in one element of the situation in step I between the deductions. Aggregative phenomena, such as equilibrium prices, are explained by adding at least two additional steps (V and VI) to the above scheme. Step V would add additional analysis about the aggregate impact of a number of agents $(A_1, A_2, \ldots, A_n)$, each doing the appropriate thing $X = (X_1, X_2, \ldots, X_n)$. The analysis in step V would take the following form: if all A_i's do X_i then the aggregate result will be Y. Step VI would then be an aggregative explanandum: therefore Y.

Notice that such an explanatory scheme really captures all of economics (at least micro), not just the textbook versions.[27] For example, the great debates over whether firms maximize profits, or satisfice, or mark up prime costs are not debates over whether the above scheme is the appropriate method of explanation. These debates are simply about what constitutes an empirically interesting specification of the situation the agent (in this case, the firm) faces. What is rigid about traditional textbook microeconomics is not that it requires adherence to the above scheme, but rather that *only certain things* are permitted in the description of the agent's situation. In particular, the Walrasian conventions of the economics profession currently seem to allow only preferences and technology as the characteristics of the agent's situation, and only prices as acceptable objective (public) constraints.

In summary then, Popper proposes situational analysis (hereafter, SA) as the only general approach for providing explanations in social science, and microeconomic explanations clearly satisfy this criterion; microeconomic explanations are special cases of SA explanations. This relationship between SA and economics raises a number of issues; some of these issues involve Popper's SA approach itself (and therefore economics), while others involve the particular form that economic explanations take *within* the general SA framework.

The most important issue concerning SA itself is that SA produces "scientific" explanations that do not satisfy Popper's own (falsificationist) criteria for scientific explanations.[28] According to Popper the falsificationist, the universal generalizations used in a scientific explanation should be scientific theories. This means, as discussed above, that such generalizations should, first, be falsifiable and, second, actually pass severe tests (be corroborated). Now consider the RP. It serves as the universal generalization in all SA explanations, it is the "law" in such explanations, and yet its nomic status is unclear.

Some claim that the RP is simply unfalsifiable; there exists no observation that would require us to give up (would logically conflict with) the claim that the agent is acting appropriately given his or her situation. It can always be argued that something is contained in (the subjective part of) the agent's situation, unknown to us, which renders the action appropriate. Others (most philosophical commentators on the issue) argue that the RP *is* falsifiable, but that it should never be abandoned, that when faced with a potentially falsifying observation we should "*cling to the rationality principle* and revise or hypothesize about his [the agent's] aims and beliefs (Watkins, 1970, p. 173).[29] In either case though, whether the RP is unfalsifiable or whether we simply choose to ignore its falsification by methodological fiat, the RP clearly is not the kind of universal generalization that Popper the falsificationist would allow in a scientific explanation. If we insist on Popper's demarcation criterion, social science explanations relying on the RP "are not *bona fide* scientific explanations" (Koertge, 1974, p. 201).[30] Or, even more strongly, since philosophers of science have traditionally considered the provision of scientific explanations to be an (possibly the) important aim of science, social science—including economics—is not science, after all. This is certainly relevant to the science of economics, but it is also relevant to the entire Popperian program since Popper explicitly developed his demarcation criteria to demarcate scientific theories in philosophy of science from those theories he considered to be pseudoscience: Marx and Freud (Popper, 1976b, pp. 41-44). These social theories can hardly be criticized for not doing what the very best social science (in Popper's view) does not do either.[31]

Before turning to the second set of issues raised by SA explanations, it should be noted that the above concerns are of fundamental philosophical importance. Some claim that to give up such explanations in social science would amount to abandoning free will in favor of complete determinism.[32] On the other hand, some claim that explanations involving RP, unlike explanations in the physical sciences, are not causal—there is no mechanism connecting the situation with the act—and thus lack a fundamental property of scientific explanations. It has been argued that this ambiguity—the ability to avoid causality and thus to preserve freedom—is precisely what makes SA explanations so attractive to Popper: "Popper wishes to escape the ugly consequences of what he considers to be Hume's dilemma by developing an account of behavior which is neither random nor determined but somewhere in between" (Latsis, 1983, p. 137).

Returning now to the more practical concerns of economic methodology, what can be said about microeconomic explanations *as* SA explanations? In other words, what if we disregard the above general criticism of SA explanations and focus on the particular form that SA explanations take in economics? What is the lesson here for economic methodology?

Sadly, the lesson is that not much can be learned from Popper's writings on SA or related work by other philosophers. If we accept SA explanations as a fact of life in social science, then all of the "action" in economics must occur in the description of the agent's situation and economists are left with all of the traditional questions regarding theory choice in their discipline. Since the RP (step III) is in *every* explanation, it is the same from one "theory" to the next. The analysis step (step II) is different for each different posited explanation; but since this second step is mostly deduction from the specifications of the agent's situation (step I), it is relatively mechanical (though it may be mathematically quite complex). It seems that the really creative part of economics—and the place where different "theories" compete for attention—is in the description of the agent's situation (step I). Economists must make decisions about how to specify the (subjective and objective) situation of the agent so that economic behavior may be predicted or explained. Recognizing that explanations in economics are all Popperian SA explanations really doesn't help with any basic issues of theory choice in economics. Economists must still make decisions about how the facts will influence their theoretical choices, how to modify the specification of the agent's situation when a prediction fails, what nonempirical considerations should influence theory choice, and so forth.

Popper and other philosophers writing on SA have focused on basically two issues. The first, discussed above, has been the question of the nomic status of the RP. The second has been the demonstration (by example) of SA success in various social sciences. Now the former is an important philosophical issue that has implications for the ultimate epistemic standing of economics, but it seems to have little to do directly with the day-to-day matters of theory choice.[33] The latter issue—no matter how psychologically satisfying positive results might be—has no real effect on the economics profession since economics is already the source of the most successful applications of SA. Thus, while economic explanations are SA explanations, and while such explanations raise important philosophical issues, the Popperian literature on SA has (at least at this time) little to offer economic methodologists concerned with the hard questions of why economists should choose one theory over another.

LAKATOS'S METHODOLOGY OF SCIENTIFIC RESEARCH PROGRAMS

Imre Lakatos's work in the philosophy of science first appeared in the early 1970s (Lakatos, 1970, 1971a) and it was adopted almost immediately by a number of people writing on economics. Numerous papers on Lakatos appeared in the economics literature, many as a result of the Nafplion Colloquium on Research Programmes in Physics and Economics held in Greece in 1974.[34] This literature on Lakatos and economics has taken basically two (non-mutually exclusive) forms. The first is mainly historical, attempting to reconstruct some particular episode in the history of economic thought along Lakatosian lines; while the second is more in the spirit of traditional work in economic methodology, attempting to appraise Lakatos's methodology of scientific research programs as an economic methodology and/or compare it to other philosophies such as those of Popper or Kuhn.[35]

In many ways Lakatos's methodology of scientific research programs (MSRP) represents a natural evolution of the Popperian tradition in the philosophy of science, but in many other respects Lakatos is quite different, addressing issues raised by more philosophers like Kuhn (1970a). For Lakatos the primary unit of appraisal in science is the "research program," rather than the scientific theory. A *research program* is a rather loose amalgam consisting of a "hard core," "positive and negative heuristics," and a "protective belt."[36] The *hard core* contains the fundamental metaphysical presuppositions of the program; it defines the program, and its elements are treated as irrefutable by the program's practitioners. To participate in the program is to accept and be guided by the program's hard core. For example, in E. Roy Weintraub's Lakatosian reconstruction of the neo-Walrasian research program in economics, the hard core consists of such propositions as "Agents have preferences over outcomes," "Agents act independently and optimize (subject to constraints)," and so on.[37] The *positive and negative heuristics,* respectively, are instructions about what should and should not be pursued in the development of the program. The positive heuristic guides researchers toward the right questions to ask and the best tools to use in answering them; the negative heuristic advises on what questions should not be pursued and what tools are inappropriate. Again using Weintraub's analysis of the neo-Walrasian program as an example, the positive heuristic contains injunctions such as "Construct theories where the agents optimize," while the negative heuristic implores researchers to avoid things like theories involving irrational behavior. Finally, the *protective belt* consists of the program's theories, auxiliary hypotheses, empirical conventions, and the (evolving) "body" of the research program. All of the activity of the program occurs in the protective belt as a result of the interaction of the hard core, the heuristics, and the program's empirical record. For the neo-Walrasian program, Weintraub argues that the protective belt includes almost all of applied microeconomics.

A research program is appraised on the basis of the theoretical activity in the protective belt. There is *theoretical progress* if each change in the protective belt is empirical content increasing, if it predicts "novel facts."[38] The research program exhibits *empirical progress* if this excess empirical content is confirmed (Lakatos, 1970, p. 118). Lakatos considers a third type of progress, *heuristic progress,* which requires the changes to be consistent with the hard core of the program. In his definitions of theoretical and empirical progress, Lakatos presupposes that conditions for this latter type of progress, heuristic progress, have been satisfied.

Lakatos's Popperian lineage is evident in a number of ways. The primary way this image is reflected is in his characterization of empirical content and novel facts. For Lakatos, like Popper, "the empirical basis of a theory is the set of its potential falsifiers: the set of those observational propositions which may disprove it" (Lakatos, 1970, p. 98, n. 2). Thus, while Lakatos clearly considers progress to be achieved through empirical confirmation, rather than falsification, his characterization of the tension between theory and fact is fundamentally falsificationist. Also with respect to empirical content, Lakatos is clearly Popperian in his "conventionalism" about the empirical basis.[39] Finally, the Popperian spirit is evident in the way Lakatos defines "metaphysical" and his recognition of the importance of metaphysics in science.[40]

On the other hand, it is safe to say that there are a number of very un-Popperian things about the MSRP. Most important of these is the complete immunity of the hard core to empirical criticism; the very idea that it is appropriate to immunize completely any part of scientific theory is in direct conflict with Popper's "nothing is sacred" falsificationist philosophy of science. Certainly Popper recognizes that science has experienced periods of Kuhnian "normal science" where the critical spirit was temporarily arrested; but for Popper this is something to lament, not praise (Popper, 1970). Another point of disagreement is the question of confirmation versus falsification. Other than the way Lakatos defines empirical content, he has little regard for falsification. For Lakatos all theories are "born refuted" (1970, pp. 120-21) and the real task of philosophy of science is to develop a method of theory appraisal that *starts* from this fact. Finally, Lakatos differs from Popper in that he embraces a historical metamethodology whereby the actual history of science can be used to appraise various methodologies of science.[41] For Popper, methodology is purely a normative affair and there is no sense in which the actual history of science can be used to "test" methodologies.

The places where Lakatos splits with Popper are where Lakatos is most likely to win the favor of economists since these places happen to be areas of substantial tension between falsificationism and economics. Certainly economics is replete with "hard cores." Not only is the rationality principle protected from refutation, but individual economic theories harbor hard-core propositions as well—Weintraub's hard-core elements of the neo-Walrasian program being an excellent case in point. While not all economists would agree on exactly what these hard core propositions should be in any particular domain of inquiry, there seems to be a consensus that such presuppositions exist and that they necessarily define alternative research programs in economics. It is not surprising that a philosophical program such as falsificationism that requires practitioners to be willing to give up any part of their research program at any instant would not be as acceptable to economists as one (such as Lakatos's) that allows for such pervasive hard cores. So too for the issue of confirmation versus falsification. It is also clear that falsificationism has not been practiced in economics and that to enforce it arbitrarily would essentially eliminate the discipline as it currently exists. On the other hand, there *is* a great amount of empirical activity in economics. The facts do matter, but they matter in a much more subtle and complex way than falsificationism allows. As Weintraub states, "the idea that facts can falsify theories, and that the role of applied work is to produce facts that falsify the theories that the theorists create, is simultaneously to misunderstand facts, theories, tests, and falsification" (1988a, p. 222). Surely Lakatos's notion of empirical progress is more like what the best empirical work in economics does, and should do, than Popperian falsificationism.

Finally, Lakatos (unlike Popper) has emphasized the role of the history of science in supporting particular methodological proposals. Of course, the question of the proper relationship between the history of science and the philosophy of science is a very complex issue that continues to be debated in philosophy, but it is clearly the case that economists have recently been sympathetic to methodological proposals sensitive to the actual history of their discipline. Economists have produced quite a vast literature that uses the Lakatosian categories to reconstruct various parts of the history of economic thought. Standard practice in such literature is to choose a particular part of economic theory (past or present) and then try to isolate and identify the hard core, the positive and negative heuristics, and the type of theoretical activity occurring in the protective belt. The bottom line in such work is usually a positive or negative appraisal of the "progressivity" of this particular part of economics.[42] Examples of such reconstructions range widely in topic area from Jevons, Menger, and Walras (in Fisher, 1986), to rational expectations macroeconomics (in Maddock, 1984), to Henry George (in Petrella, 1988).

An overall assessment of these Lakatosian case studies is very difficult: first, because of its vastness and the diversity of the literature; and second, because many economists writing in the field have taken very little care with the way the Lakatosian terminology is used. This lack of fidelity to Lakatos's concepts results in "hard cores," "heuristics," and (particularly) "novel facts" that bear little resemblance to their Lakatosian analogs or how these terms have been used in reconstructions in the physical sciences.[43] It is very difficult to evaluate such literature. Some of it is interesting (possibly creative) history of economic thought, but it is unclear what it says about the MSRP in the particular theories examined. What can be said is that, *in the case studies where the relevant language is consistent with Lakatos*, "progress"— and the prediction of novel facts it necessarily implies—has been a *rare* occurrence. There have been some well-researched cases where the prediction of novel facts has actually been uncovered,[44] but such cases correspond to a miniscule portion of the theoretical "advances" of the profession. Lakatos's criterion for theoretical progress—the prediction of novel facts—may be sufficient for what the profession considers to be theoretical progress, but it is surely not necessary. Just as "the development of economic analysis would look a dismal affair through falsificationist spectacles" (Latsis, 1976b, p. 8), it seems that economics would look almost as bad on a strict Lakatosian view. This argument assumes, of course, that we actually define such things as "progress" and "novel fact" in the Lakatosian way. If these terms are defined with sufficient vagueness (as some economists have done), then one can produce any Panglossian historical record that one desires.

Now this claim—that the MSRP has much that is relevant for economics but that empirical and theoretical advances in economics occur (and should occur) in many other ways than Lakatos specified in the MSRP—reflects very poorly (again) on Popper. The reason is that by and large *where economics is most likely to part ways with Lakatos is precisely where Lakatos borrowed most heavily from Popper.* Lakatos seems to have much to say about economics, and looking for the types of things that Lakatos suggests one should look for in the history of science has produced some important historical studies. This work has drawn attention to the discipline's metaphysical hard core; and because of its emphasis on empirical progress, it has reopened the methodological question of the relationship between applied economics/econometrics and pure economic theory. What Lakatos has not produced (and what I suspect will never be produced) is a mechanical model for the growth of scientific knowledge that provides an appropriate guide to theory choice in economics. In Lakatos's case, the fit seems to be poorest where older Popperian parts were used without much modification.

CONCLUSION

It appears that in the final evaluation "Popperian" economic methodology must be given low marks. Falsificationism—Popper's fundamental program for the growth of scientific knowledge—seems extremely ill suited to economics. Popper's situational analysis view of social science is precisely what economists do, but the discussion of the topic in the Popperian literature does not help economists with any real methodological questions. The interest in Lakatos has produced some valuable historical studies, but the overall fit of economics into the MSRP is not good—and not good precisely where Lakatos is the most Popperian.

Now, despite and even granting all of the above arguments, it could be argued that there is still a way to save the Popperian tradition from a generally negative evaluation. The defense, strongly put, is that while all of the above criticisms may be true, they really do not say anything about Popperian philosophy. The argument is that Popper's *really*

important work is something quite different from what has been discussed above, that his *real* contribution to philosophy is *critical rationalism*—not falsificationism—and that, once this is recognized, Popper still has something valuable to contribute to economic methodology.

Critical rationalism is Popper's general view of the philosophical
method. It is the general method of rational discussion and the critical
examination of proposed solutions.[45] Its overarching mandate is to
criticize, not falsify—though falsificationism is a special case of this
more general method. Falsificationism is simply critical rationalism
applied to the special case of *empirical* criticism. In addition to the
narrow empirical domain of science, critical rationalism can be applied quite generally; metaphysical theories, philosophical theories,
natural sciences that do not seem to fit falsificationism (e.g., evolutionary biology), and social sciences employing the rationality principle can all be examined, discussed, and ultimately appraised through
critical rationalism. Applying critical rationalism to economics simply
means that we should criticize economic theories and we should be
willing to learn from this critical discussion. Strategies that block or
evade criticism should be shunned, while those that open themselves
to criticism should be welcomed. If this is really Popper's position, if
this is the heart of Popperian philosophy, then the above criticisms of
falsification and Lakatos do seem to have little bearing on the overall
evaluation of the Popperian tradition and it may be that the economics
profession still has something to learn from Popperian philosophy.[46]

It is certainly difficult to argue against critical rationalism; for one
thing it seems eminently reasonable, and for another thing any rational
argument against critical rationalism seems to presuppose it. One could
argue against the exegetical claim that critical rationalism was *really*
Popper's main contribution to philosophy, but little would be gained
by doing so: critical rationalism is actually in Popper (however min-
imally), and it may be appropriate even if the claim that it was Popper's main thesis is incorrect. If there is a real problem with critical
rationalism, it is not that one can say very much against it, but rather
that one cannot say very much with it. Critical rationalism is a view
that may be palatable by virtue of its blandness—the epistemological
analog of the ethical mandate to "live the good life." Some recent
discussions of critical rationalism in the philosophical literature conclude that the notion is doomed to be a "contentless directive,"[47] too
amorphous to be of value in any interesting cases. This does not make
it wrong or pernicious—just not very informative, and devoid of the
"bite" that has been so attractive in Popperian philosophy. Thus, while
the role of Popperian philosophy can be saved by turning to critical

rationalism and away from falsification and demarcation, the victory may be relatively hollow. If one listens carefully behind the roar of such a Popperian victory speech, one may be able to hear Popper's old enemies, Hegel and Marx, chuckling in the dark.

NOTES

1. Blaug (1976a, 1987), Brown (1981), Coats (1976), Cohen (1983), Cross (1982), de Marchi (1976), Diamond (1988a), Fisher (1986), Fulton (1984), Hands (1985a), Latsis (1972, 1976b), Leijonhufvud (1976), Maddock (1984), Rizzo (1982), Schmidt (1982), and Weintraub (1985a, 1985b, 1988a)—is a partial listing of these Lakatosian reconstructions.

2. Influential books on economic method that fall broadly into the Popperian tradition include Blaug (1980a), Boland (1982, 1986), Hutchison (1938), Klant (1984), Lipsey (1966), Weintraub (1985b), and (based on Caldwell, 1988b) Caldwell (1982).

3. Popper (1976b, p. 88).

4. The term *basic statement* has a rather narrow definition in Popperian philosophy. The concept was introduced in chapter 5 of Popper (1968) and it is nicely presented in Watkins (1984, pp. 247-54).

5. Actually, as will be discussed in note 15 below (and the text), scientific theories are not *by themselves* logically falsifiable. Rather, scientific theories along with (usually numerous) auxiliary hypotheses may form logically falsifiable "test systems" (see Hausman, 1988, pp. 68-69).

6. Gellner (1974, p. 171).

7. See Hands (1988) for a general discussion of the Popperian notion of ad hocness.

8. An exception here may be Ludwig von Mises (1949, 1978).

9. See de Marchi (1988b) for a discussion of the LSE connection.

10. Popper (1961 and 1966), for example.

11. An exception here seems to be Terence Hutchison, who considers the link more direct. For Hutchison, falsificationism is "the epistemological basis for a free, pluralist society" (1976, p. 203). A similar political theme runs throughout Hutchison (1988).

12. This point is argued forcefully in Caldwell (1988b).

13. Feyerabend (1975a), Grunbaum (1976a, 1976b), Lakatos (1970), Maxwell (1972), and Putnam (1974) provide a small sample of such general criticism.

14. The main sources for this list of criticisms are Caldwell (1984a, 1988b), Hausman (1981b, 1985, 1988), Latsis (1976b), and Salanti (1987).

15. The Duhemian problem (Duhem, 1954) arises because theories are never tested alone; rather they are tested in conjunction with certain auxiliary hypotheses (including those about the data). Thus if T is the theory, the prediction of evidence e is given by $T \cdot A \Rightarrow e$, where A is the set of auxiliary hypotheses. The conjunction $T \cdot A$ forms a test system and the observation "not e" implies "not $(T \cdot A)$" rather than simply "not T"; the test system is falsified,

not necessarily the theory. In a scientific context there are a number of possible responses to "not e" (see Koertge, 1978, p. 255). One could challenge the reliability of the observation "not e." One could reject A or one of the elements of A. One could challenge the validity of the implication from T·A to e. And finally, we could follow Popper's advice and reject the theory T; discarding the theory is but one possibility.

This problem is also called the Duhem-Quine problem, though the Duhemian appellation may be exegetically incorrect (see Ariew, 1984). It is a well-known problem in the philosophy of science but has only recently been recognized as an issue for economic methodology (see Cross, 1982, for instance). Popper clearly recognized the Duhemian problem (e.g., in 1965, pp. 112, 239; and 1972, p. 353), but his methodological solution is itself subject to criticism (see particularly the discussion following the third bullet, below in the text).

16. Popper (1965, pp. 42, 267, 387-88; 1968, pp. 43-44, 93-95, 97-111; 1983, pp. 185-86).

17. Such feedback effects are related to the more traditional problem of "self-fulfilling prophecies" in social science (see Hands, 1990c).

A related problem is shown by the examples in Faulhaber and Baumol (1988). The authors discuss a number of cases where results from microeconomic theory have been applied by government and the business community. Of course, the fact that the implications of the theory were "applied" after/ because of the economic theory means that these implications did not hold when the theory was proposed. The microeconomic theories in question can be corroborated today because falsificationism was not practiced earlier. It is difficult to imagine such a case in physical science.

18. This is one source of the "innocuous falsification" mentioned by Mark Blaug (1980a, pp. 128, 259) and Alan Coddington (1975, pp. 542-45). It should be noted that if the parameters in the auxiliary hypotheses are not sufficiently restricted, but allowed to vary freely, then a different problem develops: the so-called parameter paradox (Klant, 1984, pp. 153-57; 1988, pp. 108, 110-11). This is because, if parameters are truly "variables," then *even* qualitative comparative statics cannot be obtained. This results in a theory (or theoretical test system) that is completely unfalsifiable: there exists no observation in conflict with it.

19. Popper's most important writings on verisimilitude are contained in Popper (1965 and 1972). Useful surveys of the topic are Koertge (1979b, pp. 234-38) and Watkins (1984, ch. 8).

20. This problem is demonstrated nicely by the following anecdote.

If two children are told to pick all and only the good cherries off a tree, who has done better: Clara Caution, who picks a tiny thimbleful, nearly all of which are firm and ripe, or Bella Bold, who brings home an enormous tubful, many of which are green or rotten? Which are worse, sins of omission or sins of commission? (Koertge 1979b, p. 237)

21. These concepts are discussed in detail with appropriate references to Popper's writings in Hands (1988). Other general discussions of these Popperian notions include Koertge (1978), Watkins (1978, 1984), and Worrall (1978).

22. The argument is "nowhere fully explained outside of lectures" (Jarvie, 1972, p. 51).

23. General discussions of situational analysis by philosophers other than Popper include Farr (1983), Jarvie (1972, 1982), Koertge (1974, 1975, 1979a, 1985), and Watkins (1970). The methodological discussion regarding economics is contained in Blaug (1985), Caldwell (1988b), Hands (1985b), and Latsis (1983), with brief mention in Blaug (1980a), Hutchison (1981), Klant (1984, 1988), and Latsis (1976b). Wong (1978) uses situational analysis to criticize the theoretical contribution of an individual economist, rather than examining the general implications for economic methodology.

24. Popper (1985, p. 358), Koertge (1974, p. 199), Latsis (1983, p. 136), Watkins (1970, p. 167).

25. Popper (1966, p. 97; 1976a, pp. 102-3; 1976b, pp. 117-18).

26. Koertge (1975, p. 440; 1979a, p. 87).

27. This scheme is not even restricted to "orthodox" economics. Consider the traditional Marxist answer to the question, why did nineteenth-century capitalists hire women and children and work them long hours? The answer would be that the situation of the individual capitalists was that they needed to make their rate of profit as high as possible or they would be pushed out of business and thus become proletariat themselves. The rate of profit is given by $\pi = S/(C + V)$ where S is surplus value, the amount of labor time obtained by the capitalist in excess of V, the amount of labor time necessary to reproduce the workers. Now, given C, π can be increased by increasing S or by decreasing V. Working laborers longer hours will increase S, and hiring women and children will reduce V (since V is based on reproducing the household that sustains the worker). Therefore, since agents always act "appropriately" to their situation, capitalists hired women and children and worked them long hours.

28. This issue prompted the distinction between $Popper_s$ (social science/ SA) and $Popper_n$ (natural science) in Hands (1985b) and Caldwell (1988b).

29. Popper himself certainly argues that the RP should never be abandoned (Popper, 1985, p. 360). It is unclear, though, whether this is because the RP is unfalsifiable, or whether the result of a methodological decision.

30. Actually, SA explanations are *not bona fide* scientific explanations by *any* covering law model of scientific explanations.

31. This point is made by Koertge (1979a, pp. 84, 93).

32. Latsis (1983); Koertge (1975).

33. This, of course, assumes that economics is stuck with SA. If, on the other hand, the failure of the RP as a causal law is taken seriously, one possibility would be to abandon SA explanations altogether. This would force economists to start from scratch and—if they are to explain economic behavior at all— to do their explaining on the basis of the type of causal universal laws required

in scientific explanations. This is essentially the proposal of Alexander Rosenberg (1981).

34. This conference produced the seminal volume Latsis (1976a).

35. See note 1 for references to the historical literature. The methodological literature also includes some of these same references as well as others such as Archibald (1979), Goodwin (1980), Hands (1979, 1984a, 1988), Hutchison (1976, 1981), Remenyi (1979), Robbins (1979), and Rosenberg (1986). Lakatos is also discussed in surveys such as Blaug (1980a), Caldwell (1982), and Pheby (1988).

36. Many summaries of the MSRP are available in the economics literature (Blaug, 1980a; Hands, 1985a; and Weintraub, 1985a, 1985b, and 1988a, for instance), but the single best presentation of the argument remains that of Lakatos (1970) himself. As with Popper's falsificationism, only a sketch of the main thesis is provided here.

37. As notes 1 and 35 above clearly indicate, there has been a lot of work in "Lakatosian economics." In all of this work, none has been as serious or as sustained as Weintraub's work on the neo-Walrasian program (1985a, 1985b, 1988a).

38. The term *novel fact* has been the subject of much dispute within the Lakatosian (and Popperian) program. See Gardner (1982), Hands (1985a), and Worrall (1978) on this issue.

39. This point is emphasized in Hands (1979).

40. Popper, unlike philosophers in the positivist tradition, has always recognized that metaphysics has a role to play in the growth of scientific knowledge. In fact, Popper's lifework is often characterized as a long process of systematically expanding the role of metaphysics in science (a view corroborated by the discussion of metaphysics in Popper, 1983). Philosophers in the Popperian tradition have intermittently considered the question of appraising metaphysics (Koertge, 1978; Watkins, 1958; Wisdom, 1963 and 1987, for example), but the topic remains underdeveloped. The issue will be raised again in the concluding section of this chapter.

41. "A general definition of science thus must reconstruct the acknowledgedly best gambits as 'scientific': if it fails to do so, it has to be rejected" (Lakatos, 1971a, p. 111).

42. These case studies use Lakatos to appraise economics; an exception is Hands (1985a) where economics is used to appraise Lakatos.

43. Rather than singling out the worse perpetrators of this terminological infidelity, I will take the opposite approach. In the reconstruction literature, *certain economists have been relatively careful* in the way the Lakatosian terminology is used and in the way the economic and empirical concepts are mapped into these Lakatosian notions; a list of such work would need to include Blaug (1987), de Marchi (1976), Latsis (1976b), Maddock (1984), and Weintraub (1985a, 1985b, 1988a).

44. See the references in note 43.

45. Critical rationalism has been an underlying theme throughout Popper's work. It is more pronounced in later work (esp. Popper, 1972, 1983) than earlier, but not even *The Logic of Scientific Discovery* is without it:

And yet I am quite ready to admit that there is a method which might be described as "the one method of philosophy." But it is not characteristic of philosophy alone; it is, rather, the one method of all *rational discussion*, and therefore of the natural sciences as well as philosophy. The method I have in mind is that of stating one's problem clearly and of examining its various proposed solutions *critically*. (Popper, 1968, p. 16)

46. According to Bruce Caldwell (1988b), critical rationalism is how his "pluralism" in Caldwell (1982, 1988a) should be interpreted; and it is also how Joop Klant interprets his "plausibilism" in Klant (1984)—see Klant (1988, p. 108).

47. Nola (1987, p. 479).

9

The Problem of Excess Content: Economics, Novelty, and a Long Popperian Tale

Chapter 1 was my response to the conference volume (Latsis, 1976a) from the first "Lakatos and economics" conference held in Nafplion, Greece, in 1974. This chapter is my slightly revised contribution to the second "Lakatos and Economics" conference held in Capri, Italy, in 1989 (see Hands, 1991c). The Capri conference, generously funded by Spiro Latsis, provided participants with a chance to reflect critically on the literature of the intervening fifteen years and to evaluate the overall impact of Lakatos's work on economics and economic methodology. The conference volume (Blaug and de Marchi, 1991) also contains the comments of my two discussants, Bert Hamminga (1991) and Uskali Mäki (1991a), as well as my reply (Hands, 1991a).

I have previously argued against using the Popperian/Lakatosian notions of excess content and novel facts as the sole criteria for theory appraisal in economics (Hands, 1985a, 1988, 1990a, 1992a). It is not my intention to repeat my earlier arguments here. My current task is related, but more historical. What I intend to do is to trace the sequence of events that brought Popperian philosophy (including that of Imre Lakatos) to its current position on the issues of excess content, novelty, and scientific progress. My general approach will be to analyze Popper's and Lakatos's positions on these issues as an appropriate response to the particular philosophical problem situations in which they found themselves. In particular, I will argue that Popper's concept of verisimilitude has played a fundamental role in the evolving problem situations of these two authors. Finally, after reconstructing this Popperian tale, I will return to economics and economic methodology. Some of the problems in contemporary economic methodology can be

much better understood in light of this reconstructed history of the
Popperian and Lakatosian positions.

In the first section I will reconstruct Popper's problem situation as
it pertains to the issues of content, novelty, and truth. Particular attention will be given to verisimilitude and its role in Popper's position.
The second section will discuss Lakatos's view and analyze his methodology of scientific research programs (MSRP) in the context of his
(inherited) problem situation. The third section will consider contemporary problems in Popperian philosophy—particularly problems with
the concept of verisimilitude—and relate these problems to the earlier
discussion of Popper's problem situation. The last section will, finally,
discuss economics and the relationship between these events in Popperian philosophy and recent work on economic methodology.

A LONG POPPERIAN TALE ABOUT CONTENT,
NOVELTY, AND TRUTH[1]

Popper's 1934 position in *Logik der Forschung* (hereafter, *LSD*)[2]
was purely methodological without being epistemological. His falsificationist methodology of bold conjecture and severe test provided a
set of rules for the game of science (rules for demarcating science from
nonscience, rules for correctly playing the scientific game, and a criterion for successful play or progress) without providing an ultimate
aim or purpose for playing the game. Earlier philosophies of science
had been explicit about the aim of science and had subordinated their
rules to their stated aim. "In Popper's philosophy this link seems to be
severed. The rules of the game, the methodology, stand on their own
feet; but these feet dangle in the air without philosophical support"
(Lakatos, 1978c, p. 154). Popper clearly recognized this lacuna in his
early view and sought to fill it in his later work.

Since publishing the *Logik der Forschung* (that is, since 1934) I have developed
a more systematic treatment of the problem of scientific method: I have tried to
start with some suggestions about the aims of scientific activity, and to derive
most of what I have to say about the methods of science—including many
comments about its history—from this suggestion. (Popper, 1983, p. 131)

Popper's main suggestion, introduced in 1959 or 1960, was that the
aim of science is the "search for truth."[3] "We should seek to see or
discover the most urgent problems, and we should try to solve them by
proposing true theories" (Popper, 1972, p. 44). Popper had always
preferred scientific realism and had wanted to characterize science as

the search for truth, but in the early 1930s when he was writing *LSD* the correspondence theory of truth was in such disrepute that Popper strategically chose to "avoid the topic" (1965, p. 223). It was not until Popper became familiar with Alfred Tarski's theory that he lost his "uneasiness concerning the notion of truth" (Popper, 1972, p. 320) and formally endorsed truth as the aim of science.[4]

Actually Popper went far beyond Tarski and the Tarskian notion of truth by introducing his own concept of truthlikeness or *verisimilitude*. Popper argued that if science is to aim at truth it is necessary to have a notion of approximate truth or of coming nearer to the truth. His goal in introducing the concept of verisimilitude was to be able to say "that some theory T_1 is superseded by some new theory, say T_2, because T_2 is more like the truth than T_1" (Popper, 1972, p. 47). Such a concept "allows us to say that the aim of science is truth in the sense of better approximation to truth, or greater verisimilitude" (Popper, 1972, p. 57). Popper intended verisimilitude to apply to strictly false theories; it should allow us to make sense of the notion that one theory is closer to the truth than another even if both theories are false.[5]

While an elaborate discussion of Popper's concept of verisimilitude is not required here, a brief sketch of the idea will be useful.[6] Popper's definition of verisimilitude relies heavily on the notion of the *content* of a statement (or conjecture or theory). The content of a particular statement "a," is the class of all nontautological statements logically entailed by a. This content class (call it "A") is subdivided into the truth content "(A_T)," the set of all true statements that follow from A, and the falsity content "(A_F)," the set of all false statements that follow from a. Actually we need to be a little more careful in specifying A_F since true statements can be deduced from false ones, but these definitions of A_T and A_F capture the basic ideas of truth and falsity content.[7]

Given these notions of truth content and falsity content, it is rather simple to define verisimilitude or truthlikeness. A theory T_2 has more verisimilitude than a theory T_1 if and only if a) their contents are comparable, and *either* the truth content but not the falsity content of T_2 is greater than T_1 *or* the falsity content but not the truth content of T_1 is greater than T_2. In other words, of two comparable theories, the one with more true implications (and no more false implications) or fewer false implications (and no fewer true implications) has the greater verisimilitude. Increasing this type of verisimilitude or nearness to truth became the Popperian aim of science. "To say that the aim of science is verisimilitude has considerable advantage over the perhaps simpler formulation that the aim of science is truth" (Popper, 1972, p. 57).

So, around 1960 Popper introduced the notions of truth and truth-likeness, and specified truthlikeness or verisimilitude as the aim of science. It is easy to see this move as an appropriate response to his philosophical problem situation. Popper wanted an approach that was basically realist ("the only sensible hypothesis," Popper, 1972, p. 42) while still avoiding essentialism; he perceived Tarski's notion of truth and his own concept of verisimilitude as being the solution to his problem.

> I wish to be able to say that science aims at truth in the sense of correspondence to the facts or to reality; and I also wish to say (with Einstein and other scientists) that relativity theory is—or so we conjecture—a better approximation to truth than is Newton's theory, just as the latter is a better approximation to the truth than is Kepler's theory. And I wish to be able to say these things without fearing that the concept of nearness to truth or verisimilitude is logically misconceived, or "meaningless." (Popper, 1972, p. 59)

Now, however satisfying verisimilitude might be as the aim of science, it only becomes an appropriate solution to Popper's problem situation if it does not require him to abandon the falsificationist methodology of *LSD;* not surprisingly, it does not. Popper argued that verisimilitude as an aim is perfectly consistent with the falsificationist methodology of bold conjecture and severe test that he had advocated in his earlier methodological writings.

To see this relationship, recall that for a new theory to have more verisimilitude than its comparable predecessor, the new theory needs to have either more truth content or less falsity content than the theory it replaces. First consider the case where the falsity content of the two theories is the same, and focus on the truth content. In this case, the bolder theory—the one with the most total content, the one that says more—will have more truth content and thus more verisimilitude. In this way Popper's traditional preference for bold theories, theories that take greater risks, seems to be entirely consistent with verisimilitude as the aim of science. Now consider falsity content. For more verisimilitude we want less falsity content. Popper's methodology has always been a method of severe test and attempted refutation; it is a falsificationist method. This falsificationist method of seriously attempting to refute theories and eliminating those that fail these severe tests seems entirely consistent with the aim of finding theories with low falsity content. Popper summarizes these arguments in the following way.

A theory is the bolder the greater its content. It is also the riskier: it is the more probable to start with that it will be false. We try to find its weak points, to refute it. If we fail to refute it, or if the refutations we find are at the same time also refutations of the weaker theory which was its predecessor, thus we have reason to suspect, or to conjecture, that the stronger theory has no greater falsity content than its weaker predecessor, and, therefore, that it has the greater degree of verisimilitude. (Popper, 1972, p. 53)

So, Popper argued that verisimilitude as an aim was consistent with falsificationism as a method—but this is not the end of his argument. It seems that verisimilitude as a "solution" opens up another problem. The search for truth or truthlikeness is not enough; we need "interesting truth" (Popper, 1965, p. 229). "In other words, we are not simply looking for truth, we are after interesting and enlightening truth, after theories which offer solutions to interesting *problems*. If at all possible, we are after deep theories" (Popper, 1972, p. 55). The search for ever deeper theories requires *more than* simple falsificationist practice—more than bold conjecture, severe test, and the avoidance of ad hoc defensive strategems. The method of *LSD* is consistent with the search for truth but it is not enough. Popper tells us, "I have been asked, 'What more do you want?' My answer is that there are many more things I want; or rather, which I think are required by the logic of the general problem situation in which the scientist finds himself; by the task of getting nearer to the truth. I shall confine myself here to the discussion of three such requirements" (1965, p. 241).

The first of Popper's famous three requirements is that the "new theory should proceed from some *simple, new, and powerful unifying idea*" (1965, p. 241); this is the so-called *simplicity requirement*. The second requirement is the requirement of *independent testability*. "That is to say, apart from explaining all the *explicanda* which the new theory was designed to explain, it must have new and testable consequences (preferably consequences of a *new kind*); it must lead to the prediction of phenomena which have not so far been observed" (Popper, 1965, p. 241). This second requirement is designed to eliminate ad hoc modification of a falsified theory parading as a new theory.[8]

Finally, Popper's "third requirement" is that "the theory should pass some new, and severe, tests" (1965, p. 292). This third requirement, then, is *empirical success* and it simply demands that some of the independently testable implications given by the second requirement actually be corroborated by the empirical evidence.

I contend that further progress in science would become impossible if we did not reasonably often manage to meet the third requirement; thus if the progress of science is to continue, and its rationality not to decline, we need not only successful refutations, but also positive successes. We must, that is, manage reasonably often to produce theories that entail new predictions, especially predictions of new effects, new testable consequences, suggested by the new theory and never thought of before. (Popper, 1965, p. 243)

Popper argues that this third requirement may "sound strange" (1965, p. 247) because it makes the evidence that counts in the progress of science a "partly historical idea" (p. 248). Nevertheless, this third requirement is indispensable to a science that seeks theories of ever-increasing verisimilitude.

Let us now recap this rather long Popperian tale. Though Popper's preference had always been toward a version of scientific realism, in *LSD* he advocated a falsicationist methodology without connecting it to truth.[9] Later Popper not only accepted truth as the aim of science, but proposed his own theory of verisimilitude to better characterize that aim. Since verisimilitude was consistent with his falsificationist methodology, it allowed Popper to solve his earlier problem of a methodology independent of truth. But this solution raised new problems; once verisimilitude was the stated aim, the methodology needed to be expanded. Since science progressed and approached truth by means of deeper and deeper theories, certain kinds of corroborations were of special significance. Corroborations of predictions that had "never been thought of before" (or what later came to be called "novel facts") became necessary for progress toward theories of ever greater verisimilitude. This final position became part of the inherited problem situation of Imre Lakatos to which I now turn.

LAKATOS WHIFFS THE PROBLEM

The philosophical problem situation of Imre Lakatos was in part based on the above Popperian view of scientific progress and in part based on more historicist influences such as Thomas Kuhn. Lakatos's response to these influences generated the Lakatosian rebellion that attacked the Popperian conventional wisdom (presented above) at two separate levels: the methodological level and the epistemological level. The former—the methodological criticism—is the best known part of Lakatos's work and it resulted in his own methodology: the methodology of scientific research programs (hereafter, MSRP).[10] The second part of the attack—the epistemological criticism—is less well known and it resulted in Lakatos's plea to Popper for a "whiff of inductivism."[11]

Regarding the methodological issues, Lakatos saw a number of problems with the Popperian conventional wisdom. Probably most important was the fact (apparent from contemporary history of science) that Popper's falsificationist methodology was at odds with the actual practice of great science. Lakatos advocated a "quasi-empirical approach" (1978c, p. 153); he argued that a methodology "must reconstruct the acknowledgedly best games and the most esteemed gambits as 'scientific'; if it fails to do so, it has to be rejected" (1978c, p. 145). Lakatos, among others, argued that Popper's methodology failed in this regard. In particular, Popperian falsificationism fails because it "stubbornly overestimates the immediate striking force of purely negative criticism" (1978c, p. 148). For Lakatos, actual great science has succeeded in a sea of inconsistencies and anomalies (refutations), and any adequate methodology must recognize (and rationalize) these facts. Many parts of Lakatos's MSRP—particularly the concepts of a research "program," the hard core, and the protective belts—were attempts to accommodate the actual history of science and avoid these perceived problems in Popper's methodology.

One methodological issue where Lakatos did not disagree with Popper was on the importance of predicting novel facts;[12] for Lakatos's MSRP, as for the Popperian view that he inherited,[13] progress requires novelty. On the same page (1970, p. 118) where Lakatos presents his famous definitions of "theoretical progress" ("predicts some novel, hitherto unexpected fact") and "empirical progress" ("this excess empirical content is also corroborated"), he provided the following footnote accentuating the importance of novelty.

> If I already know P_1: "Swan A is white," P_W: "All swans are white" represents no progress, because it may only lead to the discovery of such further similar facts as P_2: "Swan B is white." So-called "empirical generalizations" constitute no progress. A *new* fact must be improbable or even impossible in the light of previous knowledge. (Lakatos, 1970, p. 118n)[14]

Thus while Lakatos's MSRP was critical of much of the Popperian methodology, one aspect—the importance of novel facts—was not only accepted, but actually emphasized more than by Popper. For Popper, novel facts are a bit of an add-on since much of the empirical burden is still carried by falsification and refutation. For Lakatos—who wanted to decrease (to zero) the role of negative evidence and refutation—the entire burden of scientific progress is pushed onto the shoulders of novel facts. Lakatos's solution to his perceived methodological problem situation left him with a methodological position where "the only observational phenomena which have *any* bearing on the assessment of a research program are those which are 'novel' " (Gardner, 1982, p. 1).

In the realm of epistemology, as opposed to the realm of methodology, Lakatos is more critical of his progenitor. His criticism is based on the fact that Popper, despite his claim that verisimilitude is the aim of science, remains deeply skeptical about the certainty or the guaranteed reliability of our knowledge. Popper has argued consistently that, while we may (and should) search for truth, we never know when we find it; "even if we hit upon a true theory, we shall as a rule be merely guessing, and it may well be impossible for us to know that it *is* true" (1965, p. 225). For Popper there are "no general criteria by which we can recognize truth" (1965, p. 226). Not only is it impossible to recognize truth, but this lack of epistemological confidence extends even to our ability to know when we are getting closer to it: "My defense of the legitimacy of the idea of verisimilitude has sometimes been grossly misunderstood. In order to avoid these misunderstandings it is advisable to keep in mind my view that not only are all theories conjectural, but also all appraisals of theories, including comparisons of theories from the point of view of verisimilitude" (Popper, 1972, p. 58). And again: "We cannot justify our theories, or the belief that they are true; nor can we justify the belief that they are near to the truth" (Popper, 1983, p. 61).

Popper does believe that we have rational arguments for our theoretical preferences; we have rational reasons to suspect or conjecture that one theory has more verisimilitude than another, but we do not *know.* Recall the quote cited earlier where Popper was relating verisimilitude to boldness, saying, "We have reason to suspect, or to conjecture, that the stronger theory has no greater falsity content than its weaker predecessor" (1972, p. 53). Notice "we have reason to suspect," not "we know." Later in the same paper, Popper says virtually the same thing: "But if it passes all these tests then we may have good reason to conjecture that our theory, which as we know has greater truth content than its predecessor, may have no greater falsity content" (1972, p. 81). We "know" about truth content because that is a logical relation, but when it comes to the question of less falsity content—required for verisimilitude—Popper says only that "we may have good reason to conjecture that our theory . . . may have no greater falsity content."

Lakatos considers this lack of epistemological bite to be a major difficulty for the Popperian position. Lakatos's argument is that merely defining the aim of science as verisimilitude is not enough; one needs to be able to identify this progress when it occurs. It is not enough to say that novel corroborations "may be a good reason to conjecture" that we are getting closer to the truth; we need a firm connection between the two. The game of science needs a "truly epistemological dimension" (Zahar, 1983, p. 167). Lakatos gives Popper credit for his Tarskian turn, for introducing verisimilitude and promoting it as the aim of science—but Popper did not go far enough. Lakatos argues that after Popper modified his views it "became possible for the first time to define progress even for a sequence of false theories: such a sequence constitutes progress, if its truth-content, or, as Popper proposed, its verisimilitude (truth-content minus falsity-content) increases. But this is not enough: we have to *recognize* progress" (Lakatos, 1978c, p. 156). As John Watkins characterizes the problem, "it is one thing to have an aim for science and another to have a method by which we can, in favourable circumstances, actually pick out the hypothesis that best satisfies that aim" (1984, p. 279). For Lakatos, "Popper has not fully exploited the possibilities opened up by his Tarskian turn" (1978c, p. 159); he provided a solution in the form of his theory of verisimilitude, but then "shrank back" (ibid.) from actually solving the epistemological problem associated with his methodology.

Lakatos's solution to this epistemological problem is a "plea to Popper for a whiff of 'inductivism' " (1978c, p. 159). Lakatos recommends "an inductive principle which connects realist metaphysics with methodological appraisals, verisimilitude with corroboration, which reinterprets the rules of the 'scientific game' as a-conjectural-theory about the *signs* of the growth of *knowledge,* that is, about the signs of *growing verisimilitude of our scientific theories*" (1978c, p. 156). Lakatos argues that without such an inductive principle, any methodological proposals are mere conventions without epistemological bite.

> Without this principle Popper's "corroborations" or "refutations" and my "progress" or "degeneration" would remain mere honorific titles awarded in a pure game. With a *positive* solution of the problem of induction, however thin, methodological theories of demarcation can be turned from arbitrary conventions into rational metaphysics. (Lakatos, 1978c, p. 165).

In summary then, Lakatos's solution was to modify Popper's methodology—downplay falsification, emphasize novel facts, and add both historical and heuristic dimension to appraisal—but basically take as given Popper's aim and his characterization of verisimilitude. On the epistemological side, Lakatos was more radical; he pleaded for a "whiff of inductivism," a positive solution to the problem of induction that would connect the purported methodological rules (his or Popper's) with verisimilitude as an aim. This brings our discussion to the developments in postverisimilitude Popperian philosophy of science.

THE TALE UNRAVELS

So we have heard a long Popperian tale with a Lakatosian twist on the end—where does the story go from Lakatos? Unfortunately the story starts to unravel beyond Lakatos. I will not provide a complete discussion of these events, but let me at least review the previous discussion and comment on the evolution of the main parts of the story.

On the final issue discussed above—Lakatos's plea for a whiff of inductivism—there is relatively little to say. Lakatos's proposal, however guarded his presentation, is basically a plea for justification of the method of science in terms of the truthlikeness of the theories the method produces. However recurrent such pleas have been in the history of philosophy, it seems unlikely that such a justification will soon be upon us. Some neo-Popperians—Watkins (1984) in particular—argue that Lakatos's concern is legitimate but that the problem is not to be solved in Lakatos's way, by connecting corroborated appraisals to verisimilitude appraisals with an inductive principle (Watkins, 1984, pp. 282-88). Rather, Watkins rejects verisimilitude as the aim of science and proposes his own adequacy requirements (1984, p. 124) for any such aim. Given these broader adequacy requirements, Watkins argues that corroborations are in fact sufficient for the aim of science (1984, p. 306); this provides a possible solution to Lakatos's problem without actually responding to his "plea."

Clearly the most important factor in the unraveling of the later Popperian position has been the "admitted failure" (Popper, 1983, p. xxxv) of Popper's definition of verisimilitude. This definition, which formed the backbone of Popper's methodological proposals regarding progress and novelty, has encountered a number of serious difficulties. Ian Hacking, in a discussion of Lakatos's work, refers to verisimilitude as "Popper's hokum" (1979, p. 387), while for Joseph Agassi it is simply "a boo-boo" (1988, p. 473); and for advocates of alternative non-Popperian interpretations of verisimilitude such as Graham Oddie, Popper's work on verisimilitude has simply produced "embarrassing results" (1986, p. 164). The recognition of these difficulties initiated with two papers, Miller (1974) and Tichy (1974), which demonstrated that no false theory ever has more verisimilitude than any other false theory. Since being able to make sense of statements like "Newton's theory is closer to the truth than Kepler's theory" was one of the most important reasons for the verisimilitude concept (Popper, 1972, p. 59), these initial negative results represented a serious setback for the program. Following these 1974 papers, the critical literature has continued to such an extent that Popper now fully admits the failure of the project.

> A new definition is of interest only if it strengthens a theory. I thought that I could do this with my theory of the aims of science: the theory that science aims at truth *and* the solving of problems of explanation, that is, at theories of greater explanatory power, greater content, and greater testability. The hope further to strengthen this theory of the aims of science by the definition of verisimilitude in terms of truth and of content was, unfortunately, vain. (Popper, 1983, p. xxxvi)

While at this point, then, almost everyone admits that Popper's definition of verisimilitude has severe problems, opinions differ regarding the *importance* of these difficulties. Some, such as Elie Zahar, consider these results to be "a severe setback for fallibilist realism as a whole" (1983, p. 167). Others, such as John Worrall, merely say that "the attempt to supply realism with an 'epistemological ingredient' via the notion of verisimilitude should, perhaps, be regarded as an aberration" (1982, p. 229). Still others, such as John Watkins (1984) discussed briefly above, simply reject verisimilitude as the aim of science and build their own neo-Popperian program without it. Popper himself, of course, does not see the difficulties of his technical-quantitative definition of verisimilitude to be at all significant.

> The widely held view that scrapping this definition weakens my theory is completely baseless. I may add that I accepted the criticism of my definition within minutes of its presentation, wondering why I had not seen the mistake before; but nobody has ever shown that my theory of knowledge, which I developed at least as early as 1933 and which has been growing lustily ever since and which is much used by working scientists, is shaken in the least by this unfortunate mistaken definition, or why the idea of verisimilitude (which is not an essential part of my theory) should not be used further within my theory as an undefined concept. (Popper, 1983, pp. xxxvi-xxxvii)

Such a disclaimer is relatively easy for Popper in the 1980s since his latest turn has been toward "critical rationalism." This is the view, often associated with W. W. Bartley III, that rationally accepted propositions are those which have been critically discussed—that rationality simply means openness to criticism.[15]

Despite these Popperian disclaimers about the importance of verisimilitude, there are at least two places in Popperian philosophy where the problems of verisimilitude clearly have important (and negative) ramifications. Both of these places, by the way, are important to the relationship between Popperian philosophy and economic methodology. The first place where the failure of verisimilitude matters is in Popperian philosophy of social science, his so-called situational analysis view of social science explanations. The second place where verisimilitude matters is in Lakatosian methodology. Let us consider Popper's philosophy of social science first.

Popper's *situational analysis* approach to social science requires a *rationality principle* that serves as the law in situational analysis explanations. This rationality principle is "an integral part of every, or nearly every, testable social theory" (Popper, 1985, p. 361), and yet "the rationality principle is false" (ibid.). Now assuming that the aim of science is truth, and assuming (as Popper clearly does) that there are no fundamental methodological differences between social and physical science, the falsity of the rationality principle represents a real difficulty. But the solution to the problem is easy: verisimilitude saves the day. If verisimilitude is the aim of science and if one false theory can have more verisimilitude than another false theory, then the notion of progress toward truth need not be lost when theories involve the rationality principle.

> Ultimately, the idea of verisimilitude is most important in cases where we know that we have to work with theories which are *at best* approximations—that is to say, theories of which we actually know that they cannot be true. (This is often the case in the social sciences.) (Popper, 1965, p. 235)

> The explanations of situational logic described here are rational, theoretical reconstructions. They are oversimplified and overschematized and consequently in general *false*. Nevertheless, they can possess a considerable truth content and they can, in the strictly logical sense, be good approximations to the truth, and better than certain other testable explanations. In this sense, the logical concept of approximation to the truth is indispensable for a social science using the method of situational analysis. (Popper, 1976a, p. 103)

Notice Popper said that verisimilitude was indispensable for social science using the rationality principle.

Now consider Lakatos and the MSRP. Recall Lakatos's problem situation and how he came to advocate novel facts as the sole criterion for progress in science. Popper, in attempting to improve his basic LSD methodology so it was more in agreement with verisimilitude as the aim of science (i.e., attempting to capture the notion of ever deeper theories), proposed novel facts as a particularly significant form of corroboration. Lakatos, picking up on Popper's suggestion, abandoned falsificationism entirely, and elevated novel facts to the sole criterion for scientific progress. If Popper had never taken the Tarskian turn, never advocated verisimilitude as the aim of science, the notions of novelty and independent testability would be given only a minor role in Popperian methodology. Lakatos would have needed either to present a more traditional Popperian (i.e., falsificationist) methodology or to move entirely away from Popper, take a big whiff of inductivism, and simply appraise research programs on the basis of how well they were confirmed by the evidence. As it worked out, Lakatos could advocate novel facts as the sole criterion for progress in science and stay attached (however insecurely) to the Popperian tradition. But this linkage—so important to Lakatos—is, in light of the entire Popperian story, quite thin and frail. If what is known now about verisimilitude had been known to Lakatos, there is a very real possibility that the MSRP would characterize scientific progress in an entirely different manner.

ECONOMICS: FINALLY

This Popperian tale can give us a number of important insights into debates in contemporary economic methodology. First of all, consider Lakatos and the MSRP. Economics has been fertile ground for application of Lakatos's MSRP. Because of Lakatos's novel fact requirement for progress, many of these economic applications have amounted to novel fact hunts. And in particular, since there are now many different definitions of novel facts in the literature, these papers can become mired in semantic debates that provide little insight into either economics or philosophy.[16] More useful Lakatosian notions like the metaphysical hard core, the programmatic nature of scientific research, and the positive and negative heuristics often get neglected in the rush to find novel facts. In light of the above Popperian story, novel facts have a minor role in the overall Popperian approach to the philosophy of science; they were introduced to help forge a link between methodology and verisimilitude—which now seems either futile or unnecessary (depending on how one views the final evolution of the Popperian position). If Lakatos is going to continue to play a role in economic methodology, then in light of the above story we should reevaluate the various roles that have been played by the different parts of his position.

Popperian philosophy of social science and situational analysis is another area where the developments in Popperian philosophy (particularly verisimilitude) have an impact on economic methodology. Microeconomic explanations are a special case of situational analysis using the rationality principle. Situational analysis itself is a special case of so-called folk psychology: explanations of behavior in terms of the beliefs, desires, and intentions of the relevant agent. Recently folk psychology (and therefore, by implication, situational analysis and microeconomics) have come under attack by philosophers who argue that all such intentional explanations are not legitimate scientific explanations. While most of these critics (and defenders as well) are concerned with folk psychology in general, Alexander Rosenberg (1981, 1988a) has specifically indicted economics in this regard. Now, the success of Popper's verisimilitude program would have provided at least a partial realist defense of such explanations (and thereby micro-economics). But as it is, with the failure of verisimilitude, Popperians are left with only a critical-rationalism-as-a-default sort of argument in favor of situational analysis. This issue—the question of how (if) it is possible to regard intentional explanations as valid scientific explanations—is an important issue in the philosophy of economics. One possible defense—the one that seemed to be envisioned by Popper—has apparently collapsed with verisimilitude.

Finally, there are the important questions of instrumentalism and essentialism. Let me consider instrumentalism first. Many (perhaps most) practicing economists think of their work in instrumentalist terms. There does not seem to be any one single explanation for this instrumentalist preference. Some of the many things that might reasonably be suggested as an explanation include the nature of econometrics and the ease with which it allows prediction from past empirical trends, the demand for economic predictions from government agencies and private firms, and (possibly) the influence of Milton Friedman's instrumentalist methodology.[17] Popper, though, has always been an outspoken critic of instrumentalism.[18] His major criticism is that instrumentalism does not distinguish between science and mere technology; instrumentalism makes scientific theories "nothing but computational rules" (Popper, 1983, p. 113) and, according to the instrumentalist, "scientific theories cannot be real discoveries: they are gadgets. Science is an activity of gadget-making—glorified plumbing" (1983, p. 122). Despite arguments such as these and a sustained anti-instrumentalist rhetoric, Popper's fallibilist philosophy of science hovers very close to instrumentalism: the empirical basis of science is accepted by convention, the methodological rules themselves are merely conventions, and all knowledge is conjectural (among other things). Verisimilitude temporarily seemed to widen the gap between Popper and instrumentalism, but the failure of the verisimilitude program has again moved them closer. In Lakatos where falsification makes no contribution to science and progress occurs only through a special type of corroboration, the instrumentalist flavor seems to be even stronger.

The fact of instrumentalist practice in economics and the anti-instrumentalism of Popperian philosophy generates a tension in Popperian economic methodology. It is clearly the case that economic methodologists *need* to make sense of instrumentalism in economics, but the dominance of Popperian ideas and the negative evaluation of instrumentalism in Popperian philosophy have focused the attention of economic methodologists away from serious consideration of this instrumentalist practice.

On the other hand, Popperian philosophy is also anti-essentialist. Not only did the Popperian influence in economic methodology contribute to the neglect of instrumentalist practice, it seems to have biased the discussion against essentialism as well. While much of contemporary economic practice may appear to be instrumentalist, the founders of modern microeconomics—Menger in particular—were committed to essentialist realism.[19] Of course, this can also be said of the Marxian tradition in economics and, on a broader (non-Popperian) definition of essentialism, perhaps a significant portion of modern economics as well. It is very likely that Popper's influence has simultaneously turned our attention away from *both* how the founders of the profession perceived the discipline *and* how modern economists perceive their practice.

None of these problems might matter too much if Popper's program of conjectural realism were on firm foundations; but it is not, and it probably will not find such foundations in the near future. The failure of verisimilitude is certainly an important part of these difficulties. Lakatos's MSRP—however pregnant it might be with interesting ideas— is also unable to provide the requisite forward thrust. Economic methodology has been dominated by Popperian ideas for quite a number of years; and in light of the entire story, that domination may have lasted long enough. There are still many issues in economic methodology that could benefit from the Popperian light. But in using it, we should keep two points in mind: (1) the history of how this particular light came to shine the way it does, and (2) that it may also be useful to consider other lights.

NOTES

1. Before I even start this tale, it should be noted that many Popperian philosophers will find my argument to be too much of a Lakatosian version

(or possibly adulteration) of the story. In fact, my story *is* very close to Lakatos's version of the story, but the reason is not that I am unfamiliar with the alternatives; the reason is that I think the Lakatosian version is basically correct. The story that follows seems to rationalize the value-impregnated history of Popperian philosophy much better than any of the other stories that have been told in the literature.

2. English translation, *The Logic of Scientific Discovery* (Popper, 1968): hereafter abbreviated *LSD*.

3. Popper (1965; 1972, pp. 44-84 and 319-35; 1976b, p. 150; 1983, pp. 24-27), for example.

4. Popper (1965, pp. 223-26; 1968, p. 274n; 1972, pp. 319-21; 1976b, pp. 98-99).

5. Popper (1965, p. 235; 1972, pp. 56-57) and Koertge (1979b, p. 234).

6. Popper's primary discussion of verisimilitude is contained in chapter 2 (esp. pp. 44-60) and chapter 9 (esp. pp. 329-35) of Popper (1972) and chapter 10 (esp. pp. 231-35) and addenda (esp. pp. 391-402) of Popper (1965). The topic is surveyed in a number of works by other authors: Koertge (1979b, esp. pp. 234-38) and Watkins (1984, esp. pp. 279-88), for example.

7. The interested reader may consult Popper (1972, pp. 48-50) for a more detailed discussion.

8. Much of the above Popperian tale could be couched in "ad hoc" terms, that is, in terms of finding theories that are non-ad-hoc. In this discussion, I will try to skirt the ad hocness issue as much as possible, since I have discussed the topic in detail elsewhere (see Hands, 1988).

9. In fact in *LSD* Popper said; "I think that it would be far from *'useful'* to identify the concept of corroboration with that of truth" (1968, p. 276).

10. Lakatos (1970).

11. Lakatos (1978c).

12. An extensive literature has developed around Lakatos's use of "novel fact." Did his concept differ from Popper's? Did he have more than one concept? Can his definition of novel fact be modified so that the MSRP will be able to rationalize a greater portion of the actual history of science? I will try to avoid these controversies by simply arguing that Lakatos considered the prediction of novel facts to be necessary for progress in science, and that Lakatos's notion of novel fact was "something" like Popper's. The interested reader who would like to pursue the question of how the term *novel fact* is properly defined should examine some of the extensive literature on the topic (e.g., Carrier, 1988; Gardner, 1982; Musgrave, 1974; or Worrall, 1978).

13. As we will see in the next section, this view is not necessarily the same as the view of the most recent Popper. As my references in the last few paragraphs of the previous section indicate, the best example of the view Lakatos inherited is "Truth, Rationality, and the Growth of Knowledge," chapter 10 of Popper (1965). It is probably fair to say that this paper is the best presentation of the "middle" (verisimilitude) Popper, as opposed to the "early/*LSD*" (don't talk about truth) Popper, or the "final" (critical rationalist) Popper mentioned briefly in the next section.

14. Later (1970, pp. 155-57) in his discussion of "budding research programmes," Lakatos states that factual novelty may not be "immediately ascertainable," but he never alters his view of the fundamental role of novel facts in scientific progress.

15. This is not the place to become involved in this latest Popperian turn, but I cannot resist a passing Lakatosian quote on the topic: "the basic weakness of this position is its emptiness. There is not much point in affirming the criticizability of any position we hold without concretely specifying the forms such criticism might take" (Lakatos, 1978c, p. 144n).

16. I may have been partially responsible for some of these debates myself (Hands, 1985a).

17. See Friedman (1953).

18. See Popper (1965, ch. 3; and 1983, pp. 112-31), for instance.

19. See Mäki (1986 and 1989b).

10

Thirteen Theses on Progress
in Economic Methodology

This chapter was originally prepared for a symposium on "The State and Prospects of Economic Methodology" at the University of Helsinki in August 1989, and was subsequently published in the Finnish Economic Papers *(see Hands, 1990b). The other papers presented at the symposium were Caldwell (1990) and Mäki (1990b). The paper is rather brief, and focuses more on economic methodology in general than the Popperian tradition in particular, although many of the "thirteen theses" do in fact relate to work in the Popperian tradition.*

In the past few years there has been an explosion in the literature on economic methodology. Specialty journals like *Economics and Philosophy* have appeared, traditional economics journals have increased their coverage of methodological work, and numerous books have been published on the topic. In fact, many of these recent books are actually textbooks on economic methodology—something that indicates methodology courses are becoming a more frequent part of the academic curriculum.

Despite this increase in research and interest, many economists still claim that economic methodology is sterile, that progress never occurs, and that debates go on and on without the participants ever reaching a consensus. This lack of consensus is often cited as a reason for not participating in methodological discourse, or for disregarding those who do. Economists who raise methodological or philosophical issues often lose intellectual status roughly in proportion to the fraction of their work dedicated to such issues. As Caldwell (1990) demonstrates, there are many factors contributing to this negative attitude about economic methodology, but certainly the perceived lack of consensus is one of the more important issues.

This criticism of economic methodology is unjustified. We have learned quite a lot from the recent methodological literature, and there is now a consensus on a number of important points. In order to defend this claim, I would like to provide a list of some of the things that have been learned from recent methodological discussion. The list is comprised of things that now seem to be generally accepted by those writing on economic methodology, but that were not generally accepted by those writing on economic methodology before the late 1970s. Of course, listing thirteen theses does not demonstrate metamethodological "progress" in anything but the consensual sense, and I do not mean to imply that there is absolutely complete agreement on any one of these thirteen points. The list is only meant to show what *most* economic methodologists have learned from the past decade or so of research in the field.

The theses are not presented in any particular order of importance or degree of consensus, and the references cited are only indicative and not exhaustive of where the arguments can be found. The one exception is thesis 13, which, unlike the others, is not actually a point of consensus but rather one of my own personal suggestions that I have offered so that it might become a point of consensus in the future.

THIRTEEN THESES

1. *Econometric work does not "test" fundamental economic theory in a way that would satisfy most philosophers of science or in the way suggested by the standard rhetoric of econometric testing.* Econometric work—by far the most common type of research in economics—generally provides only a weak corroboration of economic theories. The core theory of individual maximization is never challenged by econometric evidence; and while applied theories are often disciplined by the econometric evidence, there is seldom a crucial test that causes the theory to be abandoned. In applied subfields the facts do matter, but they matter in a much more subtle and complex way than either philosophy of science or econometric rhetoric would lead us to believe. See de Marchi and Gilbert (1989), Hendry (1980), Leamer (1983), Morgan (1988).

2. *Falsificationism—the methodology of bold conjecture and severe test—is often preached in economics but it is almost never practiced.* This point requires little discussion; it is one of the most generally agreed upon of all the thirteen theses. Controversy certainly remains over whether a falsificationist methodology "should" be practiced in econom-

ics, but almost everyone agrees that it very seldom is. See Blaug (1980a), Caldwell (1984a), Hausman (1985, 1988), Salanti (1987).

3. *Though "hard cores" and "positive heuristics" abound, "novel facts" as defined by the Lakatosian school have been few and far between in the history of economic thought.* Economic methodologists seem to agree that most economic research programs do contain hard cores (metaphysical presuppositions that are taken as beyond dispute by the program's participants) and positive heuristics (guides as to what constitutes an interesting problem within the program); there continues to be disagreement over exactly what these hard core and heuristic propositions are, but there does not seem to be any disagreement over the fact that they exist and that they define alternative research programs. On the other hand, the prediction of novel facts (facts that "had never been thought of before"), which is absolutely necessary for scientific progress in the Lakatosian framework, has seldom played an important role in the development of economic theory. See Blaug (1980a, 1987), Hands (1985a, 1990a), Latsis (1976a), Weintraub (1985b, 1988a).

4. *The "Duhemian problem" is particularly difficult in economics: the complexity of economic phenomena and questions about the empirical basis of the discipline make empirical testing an extremely complex affair.* The Duhemian problem (Duhem, 1954) arises because a particular theory is never tested alone; rather, theories are always tested in conjunction with "auxiliary hypotheses" (boundary conditions, regularity conditions, ceteris paribus clauses, simplifying assumptions, etc.). This means that a negative observation only falsifies the "test system" (the theory conjoined with the auxiliary hypotheses), not necessarily the theory itself. This is a particularly problematic issue in economics, where phenomena are complex and the data questionable. See Blaug (1980a), Cross (1982), Hausman (1988), Hayek (1967).

5. *It is not the case that one particular general theory or research program in economics (neoclassical microeconomics, Keynesian macroeconomics, American institutionalism, Marxian economics, etc.) is clearly "science" while the others are clearly "nonscience."* For many years, economic methodology was used primarily to attack opponents in debates over economic theory. Economist A, who did not like Economist B's theory, would use a convenient philosophy of science to "prove" that Economist B's theory was not really "science." Seldom did anyone try to show that his or her own pet theory really was science; the general strategy was to win by disqualification of one's opponent. Hopefully this kind of "economic methodology" is a thing of the past. The nature of scientific knowledge is a subtle and complex thing. There are no simple rules that allow us to simply "accept" or "reject" any of the major eco-

nomic research programs on purely methodological grounds. There is no specific reference here, since the argument dominates most of recent methodological discussion.

6. *Milton Friedman's famous essay on economic methodology (1953) is most coherent if it is interpreted as an argument in favor of some form of instrumentalism.* It is now relatively standard to interpret Friedman as an instrumentalist, that is, holding the view that scientific theories are merely instruments. Since there are a number of different interpretations of instrumentalism, not everyone agrees on exactly which type Friedman is, but most agree that in his methodological writings he is an instrumentalist of some type. See Boland (1979), Boland and Frazer (1983), Caldwell (1980, 1991b), Hirsch and de Marchi (1984, 1990), Musgrave (1981).

7. *Discussions in economics, like discussions everywhere, have a fundamentally rhetorical component.* While most methodologists would not subscribe to the view that the tools of classical rhetorical analysis are the only tools appropriate to analyze the writing and speaking of economists, almost everyone would accept the argument that economic discourse has a fundamentally rhetorical component. See Klamer (1988a, 1988b), Mäki (1988a), McCloskey (1983, 1985, 1988a, 1988b, 1989a).

8. *The axiomatic mathematical structure of modern general equilibrium theory (the Arrow-Debreu-Walras model) does not relate to empirical economics in the same way that mathematical physics relates to empirical work in physics.* There is still a lot of methodological controversy regarding general equilibrium theory and mathematical economics, and (in particular) how they are related to applied economics and econometrics. What has been determined is that the relationship is clearly "different" from the relationship in physics. See Balzer (1982), Hamminga (1983), Handler (1980a), Hands (1985c), Hausman (1981b), Nelson (1989).

9. *Regardless of how modern theorists view neoclassical microeconomics, the founders of the theory (in the 1870s) viewed the theory from a realist perspective.* In particular, the founders of neoclassical economics were *not* instrumentalist, as many modern neoclassical economists claim to be. The two best examples of this early neoclassical realism are Carl Menger and Leon Walras: Menger was an essentialist realist, and Walras held a Platonic view. See Jaffé (1980), Koppl (1989), Mäki (1989a, 1989b), Rosenberg (1980).

10. *Neoclassical explanations of the behavior of individual economic agents are a particular form of "folk psychology" (explanations in terms of beliefs, aims, and desires). Thus, neoclassical explanations share many of the philosophical problems and/or blessings of folk*

psychology. Folk psychology—the explanation of human behavior in terms of the "intentions" of agents—has been a topic of much recent discussion in the philosophy of psychology and the philosophy of mind. Some philosophers argue that such intentional explanations cannot be legitimate scientific explanations, while others argue that such explanations are not particularly problematic. However this debate in the philosophy of mind ultimately comes out, there may be implications for microeconomics, since microeconomic explanations of individual human action are a very specific form of intentional explanation. This does not mean that microeconomics will necessarily stand or fall with folk psychology, since microeconomics is concerned with things other than individual behavior (like the unintended consequences of such behavior); but it does mean that the relationship requires careful attention. See Mäki (1989b), Nelson (1990), Rosenberg (1981, 1983, 1988a).

11. *The maximization assumption is a basic methodological presupposition of neoclassical economics. (It is not a tautology, but it is seldom falsifiable.)* The assumption that agents and firms maximize is not a tautology: it is not true simply on the basis of the definitions of the terms. On the other hand, the general proposition that something is always being maximized is seldom falsifiable by observing the behavior of agents or firms. The assumption is clearly one of the fundamental methodological presuppositions of the neoclassical research program. See Boland (1981, 1983), Caldwell (1983b).

12. *Neoclassical explanations of the behavior of agents or firms are a special case of "situational analysis."* Situational analysis explanations are explanations of an agent's behavior in terms of the agent's "situation" and the "rationality principle" that all agents act rationally given their situation. Neoclassical explanations are special cases of this type of explanation. See Caldwell (1988b), Hands (1985b), Langlois and Csontos (1989), Latsis (1983), Popper (1985).

13. *Neoclassical microeconomic theory is primarily an explanatory rather than a predictive theory, while Keynesian macroeconomic theory is primarily a predictive rather than an explanatory theory.* There are no references here; this is not, in fact, a current point of consensus. My argument is simply that the so-called symmetry thesis—the proposition that explanation and prediction are merely two sides of the same coin—does not apply in economics. What makes neoclassical microeconomics most successful is its apparent ability to provide acceptable explanations of microeconomic phenomena. On the other hand, what makes (made) Keynesian macroeconomics most successful is (was) its ability to predict the behavior of aggregated economic variables.

11

The Popperian Tradition in Economic Methodology: Should It Be Saved?

Popperian economic methodology has been subjected to some rather harsh criticism in the previous chapters. Falsificationism—to many, the real heart of the Popperian program—was criticized directly in Chapters 3 and 8 and indirectly in Chapter 2. Popper's situational analysis (SA) approach to social science—while consistent with much of what economists actually do—was found to offer little methodological guidance when it was examined in Chapters 6 and 8. Lakatos—perhaps the hardest hit—was given some guarded support in Chapter 1, but then was rather harshly criticized thereafter (particularly in Chapters 4, 5, and 8). Chapter 9 shed some rather gloomy light on the tale of Popper's efforts to come to grips with his own problem situation, and a similar tone pervaded many of the other topics considered more briefly. All in all, the discussion in the previous ten chapters has painted a rather bleak picture for Popperian economic methodology. The question to be addressed in this final chapter is this: given all these criticisms, is it still possible or desirable to save the Popperian tradition in economic methodology?

In one sense, the answer to this question is clearly *no*. If the Popperian tradition is viewed solely as *a particular normative philosophy of science*—that is, as a particular set of rather narrow *methodological rules* and the associated epistemological claims that underwrite those rules with respect to the *foundations of knowledge*—then no, the Popperian tradition cannot be saved. The arguments in the previous chapters have clearly demonstrated that, at least with respect to economics, the Popperian program has *failed in this traditional foundationalist sense*. There simply are *not* convincing arguments that by strictly following a falsificationist methodology or by restricting our attention to

research programs that consistently and correctly predict novel facts
we will necessarily be assured that our theories will produce economic
Knowledge. All of the critical remarks about Popperian economic
methodology contained in the preceding chapters are an attempt to
demonstrate—carefully, in detail, and with fidelity to the relevant
philosophical and economic texts—exactly how and why this failure
occurs.

Of course, a defender of the Popperian tradition could easily respond to these criticisms by pointing out that *no foundationalist philosophy of science is without similar difficulties*. It could easily be said
that my arguments and the arguments of others who have been critical
of Popperian methodology have not really isolated substantive difficulties within the Popperian tradition, but rather have merely reflected
certain aspects of the more general malaise within foundationalist
philosophy of science. Thus, the argument could be made that, if we
were to quit thinking of the Popperian tradition in foundationalist terms,
quit thinking of it as providing a narrow set of epistemologically justified methodological rules, and start thinking of it more (and more
properly) in postfoundationalist, critical rationalist, terms, then the
supposed "problems" with the Popperian tradition would simply *disappear.*

I am frankly uncertain how confident I am about such a Popperian
response. As will be clear from the arguments presented in the next
two sections, I certainly think that it is reasonable to offer such a defense,
and that such a defense is worthy of serious consideration. By the same
token, as will be clear from some of the notes connected with the next
two sections,[1] and by my comments in the conclusion, I also recognize
a number of potential problems with this response. What I am *not* uncertain
about is that this critical rationalist response is currently the best available
defense of the Popperian tradition: abandon foundationalist rule making but retain Popperian philosophy in modern critical rationalist guise
as a general overarching philosophical framework for methodological
discussion. This seems to me to be the only reasonable way to save the
Popperian tradition in economic methodology.

The next two sections are an attempt to present a partial articulation
of such a critical rationalist defense. These sections should thus be
read as an attempt to answer this question: if one wanted to make the
strongest possible case for saving the Popperian tradition in economic
methodology in the face of all of the criticisms, what would one say?
The main argument will be comparative: the Popperian tradition—viewed
as a general philosophical backdrop, not as a set of methodological
rules—will be compared with two other contemporary philosophical

frameworks. The argument will be that the Popperian framework has the ability to absorb the best while screening off the worst from these two alternative programs. This "defense" is similar to John Worrall's comparative defense of Popper's conjectural realism: "the main argument for conjectural realism is, as Popper discerned, negative. Its virtues only become visible when it is compared with its rivals" (Worrall, 1982, p. 230).

The first, and rather long, section below considers relativism and a number of related postmodern manifestations. Popperian philosophy is offered as a defensible intermediate position that learns from, but does not entirely surrender to, this recent discourse. The second section below focuses more narrowly on explanation in the social sciences and suggests a way for the Popperian view to extricate social science from certain difficulties that push it in the direction of eliminative naturalism. Again, the Popperian program will be defended on the basis of its ability to occupy the middle ground—in this case, the middle ground between eliminative naturalist and commonsense notions in social science. The final section returns to the more general issue of evaluating the Popperian tradition in economic methodology and suggests that even the best defense—that offered in the two preceding sections—may still face a number of difficulties.

POPPER, ECONOMICS, AND ANTIMODERNISM

It is clear that one of the dominant themes—perhaps *the* dominant theme—in late-twentieth-century intellectual life is the critique of *hegemonic modernism*. Antimodernist concerns have been raised along a wide range of disciplinary fronts and have carried a variety of appellative banners: postmodernism, antifoundationalism, radical relativism, poststructuralism, deconstruction, the end-of-philosophy critique, and the new critique of reason, to name a few. The strategies of those expressing these antimodernist concerns have also ranged quite widely: from Thomas Kuhn's (1970a) relatively localized insistence on the "incommensurability of paradigms," to Richard Rorty's (1979) more global attempt to "de-philosophize the 'conversation of mankind.' " While it is impossible to evaluate the overall impact of these wide-ranging efforts, it is clearly the case that in some fields (e.g., literary criticism and certain parts of philosophy) the impact has already been substantial and it is now starting to be felt in mainstream social science.

Given the variety and the wide range of issues raised by the anti-

modernist critique, let me, in the interest of manageability, try to isolate a few aspects of the general movement and examine some of these
issues on an individual basis. Five such individual topic areas are
introduced in categories a through e below. The names attached to each
of these areas are *my own*. I have tried to be consistent with the terminology in the literature, but the terms in no way represent a consensus of usage. These five areas, while quite broad, were chosen solely
on the basis of their relevance to Popperian economic methodology
and do not constitute a comprehensive overview of the antimodernist
movement. It must be admitted that my classification is an attempt to
take an amorphous and multifaceted intellectual movement and force
it into a rather narrow but more manageable set of discrete categories.[2]
Following categories a through e is a corresponding list of categories
a' through e' that shines a Popperian light on each of these five areas.

a. *Relativism:* Since relativism comes in a variety of different forms
and can mean a number of different things (ontological relativism, cultural relativism, moral relativism, radical relativism, etc.), I would like
to narrow the focus and give the term a rather restricted definition. When
relativism is used in the following discussion, it will mean *relativism in
the philosophy of science*. According to this position, empiricist-based
philosophy of science has not been able to demonstrate that there are
cognitively justified ways of choosing between scientific theories. This
is the position that, with respect to science, "we are never warrantedly
in a position to assert that any theory is objectively superior to another,
or that the evidence and the arguments in favor of one theory are stronger
than those favoring any other" (Laudan, 1988, p. 117). This is the relativism of failed philosophy of science. It may be that science actually
gets things right; but whether it does or not is independent of the various
stories that philosophies of science have told about how science works
or how it should work. It is important to note that relativism, as it is used
here, does not disturb the traditional epistemological categories of "knowledge" and "the world." It simply doubts whether it is possible to demonstrate, as traditional confirmationist and falsificationist philosophers
of science have tried to demonstrate, that scientific theories produced by
following the traditionally prescribed methods will actually give us such
knowledge.

One source of relativism was discussed in Chapter 3: the relativism
that originates with the work of Thomas Kuhn, Norwood R. Hanson, and
Paul Feyerabend. In brief, the argument is that, if all "observations" are
theory laden and each paradigm generates the "facts" within its domain,
then different scientific theories are fundamentally "incommensurable."

Given this incommensurability, "the empirical evidence" cannot be the basis for rational choice between theories, the traditional theory-choice criteria from philosophy of science can no longer be applied, and relativism is the result.[3] As discussed in Chapter 3, this relativism—or finding ways around it—motivated many of the developments within the philosophy of science during the past three decades (including Lakatos's work), and these changes have in turn influenced the methodological discourse in economics.

b. *Postepistemology:* For the purposes of this chapter, the term *postepistemology* will mean a view, unlike relativism, that definitely does disturb the traditional epistemological categories: the *critical part* of Richard Rorty's position in *Philosophy and the Mirror of Nature* (1979).[4] The "critical part" is emphasized because Rorty's purpose was not only the deconstruction of Philosophy (with a capital P)—that is, epistemology[5]—but also the reconstruction of his own alternative to it. This alternative combines elements of pragmatism and hermeneutics to construct a postepistemological philosophy (note the lowercase p) to replace the Philosophy destroyed by his critique.[6]

The main thrust of Rorty's argument is that philosophy has long been dominated by the metaphor that the *mind mirrors the external world*. If the mind is a mirror for the world, then all questions regarding "knowledge" can simply be reduced to questions about the accuracy of the representations in the mirror. Correspondingly, on the basis of the mirror metaphor, the theory of knowledge becomes an investigation into the source of accurate representations, and the *foundations* of knowledge become merely a special class of *privileged representations*. The role of philosophy as it has evolved in such a context is to adjudicate over all branches of inquiry with respect to these epistemic foundations—to serve, in Jurgen Habermas's phrase, as the "usher" for intellectual life (1987a, p. 298). When the metaphor of mind as mirror is abandoned, as Rorty recommends, philosophy loses its image as "a discipline which stands apart from empirical inquiry and explains the relevance of the results of such inquiry to the rest of culture" (1979, p. 220). Rorty's critique urges us to "drop the notion of the philosophers as knowing something about knowing which nobody else knows so well" and therefore to "drop the notion that his voice always has an overriding claim on the attention of the other participants in the conversation" (1979, p. 392). The success of Rorty's program would end, in Stanley Fish's words, "the scam that philosophy as a discipline has been selling us for so long" (1988, p. 28).

Rorty's project of unmasking epistemology has clearly influenced discourse in contemporary philosophy: "after Philosophy" and "the end of Philosophy" are increasingly frequent topics for conferences and books.[7] However, with two exceptions, this literature has thus far had only a minor impact on economic methodology. The first exception is the literature on the "rhetoric of economics" discussed in category d below; both of the main contributors to the rhetoric of economics literature, Arjo Klamer and Donald McCloskey, claim to draw some of their inspiration from Rorty's work. The second exception is the incipient literature on "postmodernism and economics," a literature discussed briefly at the end of the next category.[8]

   c. *Radical Postmodernism:* "Postmodern" may come to be the intellectual portmanteau word of our age; it is used to describe everything from architectural styles to ways of reading. Because of this wide range of usage and because many of these usages can be quite emotionally charged, I have not listed postmodernism as one of the variants of antimodernism, though, in some sense, postmodernism *is* simply antimodernism. The way that postmodernism will be used in the following discussion will be in a rather extreme form: radical postmodernism. By "radical postmodernism" I mean the limiting case of antimodernist discourse: the position that we are forever without first discourse, the position that we must abandon entirely the notion of a privileged metadiscourse that grounds and legitimizes other discourses. Radical postmodernism is, in Jean-Francois Lyotard's phrase, "incredulity toward metanarratives" (1987, p. 74). It is the leveling and fragmentation of all discourses, all language games; it is a totalizing critique—a deconstruction—of Reason as we have come to accept it since the Enlightenment. It is the end of "arrogant reason"—reason that claims to ground everything even itself—and the abandonment of the ancient distinction between "knowledge" (*episteme*) and "opinion" (*doxa*). "In Wittgensteinian terms, it consists in the thesis that all language games coexist at the same level, side by side, without there being any possibility of providing a general theory of language games or of granting a more basic status to one of them vis a vis the others" (Dascal, 1989, p. 221).[9]
   While this totalizing critique of reason has yet to have a major influence on economic discourse, such ideas—sometimes implicitly and sometimes more explicitly—have recently started to appear in the economics literature.[10] At this point, it is unclear how much of this recent "postmodernism in economics" actually owes its inspiration to the extreme form of postmodernism defined here and how much it owes to other influences such as Rorty and the ideas discussed in categories d and e below. Radical postmodernism in the sense it is used here is introduced primarily to frame the ideas of relativism category a and postepistemology catego-

ry b—ideas that have had a clear and direct influence on the methodological discussion in economics.

Before moving on to the final two (more narrow and possibly less "antimodernist") categories on the list, it is useful to clarify the relationship between the three different ideas *relativism, postepistemology,* and *radical postmodernism* as they are being used here. In a nutshell, *relativism* (category a) says that *science* is not privileged, *postepistemology* (category b) says that *Philosophy* (particularly epistemology) is not privileged, while *radical postmodernism* (category c) says *nothing* is privileged. Alternatively, all three refer to a type of "relativism" broadly defined: category a is relativism with respect to science, category b is relativism with respect to Philosophy, while category c) is relativism with respect to the broad categories of Western intellectual life.

d. *The Rhetoric of Economics:* Suppose that one is interested in a particular discipline—say, economics—and suppose further that one subscribes to the postepistemological position of abandoning the notion of privileged representation as a guide to successful activity within that discipline. Given that truth as correspondence or accuracy of theoretical representation would not have any natural right to guide disciplinary discourse, what, if anything, would provide such guidance? One answer is simply *persuasion;* what would guide disciplinary discourse is whatever had in fact persuaded members of the disciplinary language community that it should guide. Now it is important to note that such persuasion does not necessarily rule out epistemological arguments or empirical arguments. Theories that have passed severe empirical tests and claims regarding the usefulness of such tests can be quite persuasive. It only means that such arguments give up any natural right to usher or legitimize the process of inquiry. A democratic populace may vote for their ex-king as leader—but even if the person and the day-to-day tasks of the job are the same, things are different. The issue is the elimination of (ostensibly cognitive) privilege, not the predetermination of exactly what will replace it.

If persuasion is the issue and one is interested in investigating what goes on in a particular discipline, then one must examine what that particular language community finds persuasive. One must examine the *rhetoric* of the discipline. If the discipline in question is economics, one must investigate the *rhetoric of economics.* The word *rhetoric* in this sense does not mean the "verbal shell game" of "mere rhetoric," but rather rhetoric as "the art of speaking" or "the study of how people persuade" (McCloskey, 1985, p. 29). The rhetoric of economics is thus an investigation into how economists persuade; it is a conversation about the conversation of economists.

> To call economics "rhetorical" is not to attack science. All writings with designs on its readers is rhetorical. A mathematical proof has a rhetoric, which is to say a strategy of persuasion. . . . "Rhetoric" means here the whole art of argument; it does not mean ornament or hot air alone. (Klamer and McCloskey, 1989, pp. 140-41)

The "rhetoric of economics" in this sense was introduced by Donald McCloskey in his influential paper "The Rhetoric of Economics" (1983). Since that initial paper, the topic has grown and generated a substantial literature.[11] A portion of this literature is devoted to case studies in the rhetoric of individual subfields within economics—either as current economic theory or in the history of economic thought. As noted above, this rhetorical approach *need not* exclude the consideration of epistemological discourse, but it certainly does shake up the traditional categories of inquiry.

> [The] old talk about the logic of appraisal, the criteria for truth, the logic of explanation, and the rational reconstruction of research programs is to be expanded into talk about genres, arguments, metaphors, implied authors, and domains of discourse. The point is to figure out why some arguments work in economics and others don't. Figuring it out in literary ways will resolve some of the puzzles. (Klamer and McCloskey, 1988, p. 11)

e. *The Sociology of Scientific Knowledge:* The sociology of scientific knowledge is probably the least homogeneous and the fastest growing category of the five antimodernist categories considered. Let me introduce it by considering the same question that was used to introduce the above discussion of rhetoric. Suppose that one is interested in a particular scientific discipline, but one either accepts a Rorty-type critique of epistemology or one simply wants to investigate the "beliefs" that scientists hold, independently of any question about the cognitive value these beliefs might have. Since correspondence and representation are not relevant to theory choice in this case, what is relevant? What is it that determines the theoretical choices and the practical decisions that scientists make?

The *sociology of knowledge* answers this question with the response, *"Look to social science!"* Since the scientific community (or any particular subdiscipline within it) is simply a community—a society, with its own system of beliefs, habits, customs, and rituals—then one should use the same tools to investigate the activity that goes on in science as one would use to investigate the activity in any other social community. All societies, all cultural communities, are characterized by a system of collectively held beliefs; in science, this system of collectively held beliefs is called *scientific knowledge*. The question of why scientific knowledge is what it is amounts to generally the same type of question as why any other society comes to hold the beliefs it does. Such questions have traditionally been the domain of the social sciences.

Given the many approaches to social theorizing, one would expect to find a wide variety of different approaches to the sociology of knowledge; and that is, in fact, what one finds. The surprising fact is that one also finds a number of different points of agreement among these various views. Probably the most important point of agreement among the various studies in the sociology of knowledge is that they find *little evidence that practicing scientists behave in a way that is consistent with traditional epistemological virtue* (Popperian or otherwise). As was true in the case of Thomas Kuhn and the historians of science discussed in earlier chapters, practicing scientists evidently believe their scientific theories for a number of different reasons, but these reasons are almost never the reasons that traditional philosophy of science would call the cognitive status of the theories involved. The general impact of the sociology of knowledge has been the demystification (or perhaps debunking) of the cognitive status of science, rather than its reaccreditation. Robert Nola summarizes the situation in the following way.

> In general, many sociologists and some historians and philosophers of science have regarded science as a cultural effusion essentially no different from any belief systems and practices with which it has been traditionally contrasted, such as religion or myth. They have also alleged that science has no privileged status as a means of gathering knowledge of the world, as many philosophers have claimed, and have insisted that it ought to be studied in the same way as one would study any cultural phenomenon. Just as anthropologists have recorded the beliefs and practices of alien tribes of people and have drawn relativist conclusions with respect to both beliefs and values, so sociologists, along with their fellow-travelling historians and philosophers, have entered the domains of tribes of scientists to study their beliefs and practices and have, similarly, drawn relativist conclusions with regard to their beliefs, norms and goals. (Nola, 1988, p. 2).[12]

Despite the agreement on a few fundamental issues, there still remains a wide range of perspectives within the sociology of scientific knowledge.[13] For the purposes of this chapter, though, I would like to focus on just two "schools of thought."[14] The first of these is the Edinburgh school or the so-called *strong program* in the sociology of scientific knowledge; it is associated primarily with the work of Barry Barnes, David Bloor, and others. This first group is actually cohesive enough that the term *school* is a proper label for their work. The second group is much less cohesive; it is the ethnographic or *constructivist program* associated with Harry Collins, Karin Knorr-Cetina, Bruno Latour, Steve Woolgar, and others.[15] There are at least two basic sets of differences that separate these schools of thought: one has to do with their differing views on the proper level of theoretical abstraction in social science, while the other difference concerns epistemology.

First, regarding social science, the two schools differ regarding the proper level of sociological/anthropological abstraction that should be applied to science. The *strong program* frames its investigation primarily in *macrosocial* terms, while the *constructivist* school focuses more on *microsocial* factors. For the strong program, the scientific community is embedded in a larger social context, and that larger social context—its class relations, power relations, relations of production, and social interests—determines the content of scientific knowledge. Much like the relationship between superstructure and base in Marxian analysis, the strong program considers scientific knowledge—a particular type of social belief—to be a systematic reflection of the characteristics of the more fundamental macrosocial structure. For the constructivist program, social context also matters, but it is the more localized microsocial context of the laboratory and professional life within the discipline of science that is influential, rather than the wider social, political, and economic interests that are emphasized by the strong program. For the constructivist program, scientific knowledge is socially "constructed" in the laboratory or other scientific context, and it is this socially conditioned character of scientific knowledge that requires emphasis. Since there are many microsocial factors that could influence the activities of scientists, constructivist studies tend to be carefully detailed, densely textured, and quite particularized narratives about the influence of these many factors.[16].

Second, another difference between the strong program and the constructivist approach pertains to their view about the nature of the knowledge that one obtains when one actually does the sociology of knowledge. The *strong program* is basically *empiricist and inductivist:* one attempts to uncover the social causes of scientific beliefs by observing science.[17] This is not to say that rigid methodological "rules" are in effect. Advocates of the strong program do not require rules of inductive inference that relate a particular number of empirical observations of actual science to the causes of scientific belief, but their approach is inductive nonetheless. David Bloor claims that his "suggestion is simply that we transfer the instincts we have acquired in the laboratory to the study of knowledge itself" (1984, p. 83).[18] Thus, as Uskali Mäki has recently characterized the strong program, it is "radically pro-science" and "based on a naturalistic methodological monism" (1992, p. 4).

Constructivist sociology of knowledge, on the other hand, is more *hermeneutic* in its attempt to understand scientific knowledge and the social context that determines it. The method is more anthropological, more participant-observer, and more based on the sociologist-of-knowledge's coming to understand science by being immersed in its context-laden practice.[19] Knorr-Cetina explains:

> I want to propose that we see scientific method as a heavily textured phenomenon rather than as the mere execution of some philosophically intuited standard of reason. Context, or in a broad sense, "culture" is inside the *epistemic*, and the sociology of knowledge, or perhaps we should rather say the study of knowledge, must also concern itself with the *cultural structure of scientific methodology*.
> . . . To pursue such a goal one must move inside the epistemic space within which scientists work and identify the tools and devices which they use in their "truth"-finding navigation. (Knorr-Cetina, 1991, p. 107)

Given the differences between these two schools of thought within the sociology of knowledge, and given the above categories a through c of relativism, postepistemology, and radical postmodernism, one might be tempted to associate the strong program with relativism and the constructivist approach with postepistemology. While arguments could be made for such a characterization, it is a temptation that I believe should be resisted. Things are just not that simple, and these categories are much too rough-hewn to be used in this way.[20] The categories a through e are generally quite useful, but they admittedly represent a rather crude dissection of the antimodernist literature and its related themes.

Having completed this dissection of antimodernism, let me now return to the main issue of Popperian philosophy and economics. Categories a' through e' will offer what I believe to be the best Popperian-based response to the issues raised in each of my five topic areas.

a'. First recall the definition of "relativism" from category a above: it was the position that science is "never warrantedly in a position to assert that any theory is objectively superior to another" (Laudan, 1988, p. 117). Recall that as relativism is used here it does not question the existence of the world or the possibility of our knowledge; it only doubts our ability to prove that empirical science gives us such knowledge. In other words, relativism is the position that we are never justified in asserting that one scientific theory is better than another. In defense of Popperian philosophy, it can be argued that *relativism so defined is not inconsistent with Popperian philosophy.* In fact, in his more recent work, this is exactly what Popper asserts. In the *Postscript to the Logic of Scientific Discovery* (1982, 1983), Popper makes it quite clear that he subscribes to the basic tenets of W. W. Bartley III's interpretation of his work,[21,22] and one of Bartley's main arguments is that Popper's philosophy is fundamentally nonjustificationist. According to this argument, the mainstream of Western philosophy has been dominated by justificationism (or actually foundationalism),[23] the position that questions of rationality and reason are "concerned with how to justify—i.e., verify, confirm, make firmer, strengthen, validate, make certain, show to be certain, . . . whatever action or opinion is under consideration" (Bartley, 1987a, p. 206). For Bartley, Popper's *unique* and most important philosophical contribution is that he provided a way of discussing rationality—and thus ways of making rational choices between scientific theories—that is fundamentally nonjustificationist. Popperian philosophy

> denies that justifications must be given for something to be rational; and it does not turn to description when justification proves impossible. Rather, *it abandons all justification whatsoever.* . . . While agreeing with proponents of limited rationality that principles and standards of rationality, or frameworks and ways of life, cannot be justified rationally, we regard this as a triviality rather than as an indication of the limits of rationality. For we contend that nothing at all can be justified rationally. (Bartley, 1987a, pp. 211-12)

Or, as Popper himself says, "I assert (differing, Bartley contends, from all previous rationalists except perhaps those who were driven into skepticism) that we cannot give any positive justification or any positive reason for our theories and our beliefs" (1983, p. 19). One of Bartley's main arguments in favor of an antijustificationist view is the problem of the theory-ladenness of observations and the corresponding "conventional" aspect of the empirical basis—an issue consistently emphasized by the Popperian tradition (see Chapter 1 of this book).[24]

So it can be argued that Popper is actually a *relativist with respect to the justification* of theories or beliefs, but this does *not* mean that Popper would accept the standard corollary to such justificational relativism: *relativism with respect to the rationality* of choosing certain theories or beliefs. Popper rejects the idea that theories can be justified, but he does believe that they can be rationally preferred.

According to the Bartley and senior Popper view, called "pancritical rationalism" by Bartley and "critical rationalism" by Popper, there *are* rational reasons for believing in one theory rather than another; this rationality is located in criticism.

> We can often give reasons for regarding one theory as preferable to another. They consist in pointing out that, and how, one theory has hitherto withstood criticism better than another. (Popper, 1983, p. 20)

> A rationalist is, for us, one who holds *all* his positions—including standards, goals, decisions, criteria, authorities, and *especially* his own most fundamental framework or way of life—open to criticism. He withholds nothing from examination and review. (Bartley, 1987a, p. 212)

According to *critical rationalism* (hereafter, CR), the (insoluble) problem of justification is replaced by the (far more tractable) problem of criticism; rationality is saved from relativism by *hinging rationality on criticism rather than justification.*[25]

Within the CR framework, methodological questions take on a different complexion. The question is no longer one of finding the correct methodological rules to follow to obtain scientific knowledge; rather it is changed to the question of finding a way to arrange our lives and social institutions in a way that maximizes productive criticism. As we will see below, Bartley frequently considers such problems—problems in the organization of our educational and research institutions—to be economic problems. For him they are simply problems in the industrial organization of knowledge.

This redefined methodological project of finding the optimal critical environment is clearly not restricted to science; but within the context of empirical science, Bartley's recommendations do not differ greatly from Popper's traditional recommendations of bold conjecture and severe test. Although this reendorsement of Popper's methodological prescriptions may not—given Popper's support of CR—be very surprising, it does reframe these prescriptions in at least two fundamental ways. First, it makes quite clear the observation that Popper's general method of conjectures and refutations is consistent with a particular form of relativism—call it "justificational relativism"—and therefore compatible with certain antimodernist arguments. Second, while compatible, the method is no longer "the method" in some absolute sense. Again we can make the analogy of electing an ex-king: bold conjecture and severe test may still be recommended in science, but only to the extent that such a practice would serve (and in tandem with other practices that also serve) the more general goal of further-ing critical discussion. *It means the end of arrogant falsificationism.*

Popper, throughout his career, has "leaned against the wind" with respect to the issues of justification and relativism within the philos-ophy of science. In the 1930s and 1940s when positivist ideas were most influential, Popper leaned against the empiricist foundationalism of the positivist program and emphasized the theory-ladenness and inherently conjectural nature of science's empirical basis. On the other hand, in the 1970s when the relativist ideas of Thomas Kuhn and Paul Feyerabend were becoming influential in the philosophy of science, Popper leaned the other way toward severe empirical testing as the way to judge the worthiness of scientific theories. Popper's lean in both directions can be defended. No doubt, positivism overstated the empirical foundation of our knowledge. Our scientific knowledge is not built up, one absolutely certain brick after another, from an empir-ical basis of incorrigible sense data. By the same token, the facts do matter; we make choices between scientific theories on the basis of their empirical success, and it is clearly rational that we make our theoretical choices in this way. It can be argued that the Bartley-Pop-per program of CR provides a rationale for both of these "leans." It correctly leans against justificationism but without forcing us to sur-render to the "anything goes" of relativism.[26]

b'. In many ways Popper is clearly a member of the general epis-temological tradition that Rorty intends to overthrow with his critique of epistemology. Popper is an ontological or *metaphysical realist*;[27] there is a world "out there" that exists independently of any human knowledge about it.[28] Popper also considers the goal of science to be *truth*, and truth is characterized as *accurate representation* or *correspondence.*[29]

Despite the appearance of Popper's seduction by the mirror metaphor, his entrapment is far from complete. It is possible to characterize Popper in such a way that he does not fit squarely into the epistemological tra-dition that Rorty tries to refute. Popper does, contrary to Rorty, subscribe to the view of philosophy as first discourse—philosophy as a metaframe-work that "ushers" the rest of life (or at least the parts of life involving knowledge)—but it does not usher by having the key to certainty. Pop-per's position is different; he is a *fallibilist*. We seek truth as correspon-dence or representation, but we are never certain that we have found it. For Popper, truth is "a regulative idea"; it is "a standard of which we may fall short" (Popper, 1983, p. 26), but it is never something we are certain that we have obtained. Likewise, Popper's realism is only *conjectural realism;* we are never certain that the terms in our scientific theories refer to what is actually "out there." Although we strive for theories that correspond to the world, they are forever conjectural, forever capable of (and likely to be) overthrown.[30] Now, as I argued in Chapter 9, all is not well with the conjectural realist program.[31] However, a defense of the Popperian tradition with respect to Rorty does not require that conjectural realism be without problems, buy only that conjectural-realism-cum-fallibilism should place Popper outside the range of Rorty's sweeping critique. While dodging Rorty's bullet may not seem like a very strong defense of Popper, it is in fact quite strong, given the direct hit that Rorty has scored on so many other philosophies of science.

Rorty is really concerned with justification (or foundationalism); and he argues that, given the failure of justificationism, the only alternative is his own pragmatist-hermeneutic approach. Much of his argument in *Philosophy and the Mirror of Nature* (Rorty, 1979) is to demonstrate that his own alternative actually incorporates the best from a number of traditions, including the pragmatism of John Dewey and W. V. O. Quine, and the hermeneutics of Hans-Georg Gadamer. Rorty's entire discussion is couched in terms of only one alternative (his) to the errors of the epistemological tradition. Regardless of whether Rorty's own alternative is or is not successful, Popper's conjectural realism— particularly on the CR interpretation—represents a *second alternative*. This point is made forcefully by Peter Munz.

> The second alternative was proposed by Karl Popper nearly half a century ago. Popper argued that though knowledge cannot be *justified*—not by induction and not by a bucket theory of the mind and not by a mirror metaphor—we nevertheless have knowledge. Since we obviously *do* have some knowledge and since it cannot be justified, Popper showed that the truth of knowledge does not depend on our ability to justify it. . . . It is totally incomprehensible how Rorty could have written in the seventies of this century and succeeded in completely ignoring this Popperian alternative to the justification of knowledge by the mirror metaphor. . . . By showing that we can have knowledge without "foundation" or induction or mirror images, Popper has filled the vacuum left by the demise of foundationalist epistemology. (Munz, 1987, p. 371)

Thus the argument can be made that it is possible to accept some parts of Rorty's critique—the part about the failure of foundationalism and the part about the failure of the traditional grounding of philosophical privilege—without abandoning the Popperian program. In fact, Popperian philosophy in modern CR guise represents an alternative and possibly superior response to precisely the criticisms that Rorty has so effectively raised.[32] This clearly has implications for Rorty-inspired discussion in the philosophy of science and economic methodology.

    c'. There may be very little that one can say in response to the totalizing critique of radical postmodernism. After all, if one can find a place to stand or a position from which to stage a counterattack, then one has evidently failed to understand the message.[33] Of course it could be argued that, given the extreme way radical postmodernism has been characterized, failure to provide a response may not be very significant. Radical postmodernism as the limiting case of antimodernism is such an extreme view that Rorty is probably correct when he says: "the view that every belief on a certain topic, or perhaps about *any* topic, is as good as every other. No one holds this view" (1982a, p. 166). Given the narrowness of my definition of radical postmodernism, and given that the current focus is the defense of Popperian economic methodology, it may be reasonable simply to skip category c and move on to the questions of rhetoric and the sociology of scientific knowledge, where the influence on economic methodology is more direct. Before moving on, though, I would like to mention briefly one point of contact between the Popperian tradition, as it has been characterized in the preceding discusson, and certain aspects of radical postmodernism.

There are a number of authors outside the Popperian tradition who take radical postmodernism—and, more importantly, the criticism of this position—to be a very serious task. One example is Jürgen Habermas.[34] Habermas is concerned with the "new critique of reason" (1987b, p. 302)—the postmodernist literature of Derrida, Foucault, Lyotard, and to a lesser extent Rorty. For Habermas such views are inherently *conservative* and inhibit progressive social change. The universal modernist ideals of the Enlightenment—truth, justice, and beauty—or, more particularly, the failure of these ideals to be realized can form the basis for social criticism, a potentially liberating criticism. Radical postmodernism—the absolute leveling of all discourses—levels these ideals as well, thus eliminating the critical counterdiscourse they generate: "if thinking can no longer operate in the element of truth, or of validity claims in general, contradiction and criticism lose their meaning" (Habermas, 1987b, p. 124). Habermas's own solution to this problem—his attempt to preserve the counterdiscourse of the Enlightenment—is his theory of "communicative action" a standard for evaluative judgements reached through the consensus of an "ideal speech situation."

So while Habermas is concerned with criticizing radical postmodernism, what does his view of the problem have to do with Popperian philosophy or economics? Actually it does not have much to do directly with economics, but there is an overlap with Popperian critical rationalism. Habermas considers Popperian CR—particularly CR coming by way of its German advocate, Hans Albert (1985)—to be an alternative to his own solution to the problem of saving reason.

On the face of it, the critical-rationalist position breaks completely with transcendentalism. It holds that the three horns of the "Munchhausen trilemma"— logical circularity, infinite regress, and recourse to absolute certitude—can only be avoided if one gives up any hope of grounding or justifying whatsoever. Here the notion of justification is being dislodged in favor of the concept of critical testing, which becomes the critical rationalist's equivalent for justification. (Habermas, 1987a, p. 302)[35]

Habermas even suggests the possibility of CR and the negative dialectic approach of Theodor Adorno "supplementing each other" (1987a, p. 303). Such arguments do not directly relate to economic methodology, but they do lend additional support to the central thesis of the preceding two categories: that the Popperian philosophical tradition, particularly in modern CR form, provides a reasoned response to certain antimodernist trends. Popperian philosophy does not want to throw the Enlightenment baby of reason out with the foundationalist bathwater— but it clearly does not have its philosophical head in the sand, either.

d'. There are at least two separate versions of the "rhetoric of" literature: a "thick rhetoric" and a "thin rhetoric." On the basis of the first (thick) version, the "rhetoric of economics" or the "rhetoric of" any other domain of professional discourse is really a rather diffident and unassuming project.[36] In this view, "the rhetoric of" starts from the simple fact that, in their intercourse with other professionals, people try to persuade. Ultimately the professional discourse of members of the group, their collective conversation—including their theories, "results," schools of thought, background knowledge, empirical basis, presuppositions, and so on—is a product of this persuasion. What the group members find convincing, how they argue, who they believe, and the jokes they tell— such things reveal the discursive practices of the discipline, they individuate and characterize this community relative to other professional communities, and they help us better understand the disciplinary conversation. What makes this "rhetoric of" project—the thick rhetoric project—different from a more assuming thin "rhetoric of" project is that persuasion is only one aspect of the story. In this view, what does and what does not persuade is inexorably intertwined with a number of other factors and influences. What the discipline is (its discourse and its artifacts) is a product of a number of different factors: the social relations within science (personal relations, dominance relations, gender relations, etc.), the wider social/political/economic/ideological context, the psychological propensities of the individual members of the scientific community, and the "external world," to name just a few. The external world influences the discourse of the community both internally (as when something unanticipated appears in a test tube or when inflation reaches double digits) as well as externally (when things have an impact on the wider community in which science is embedded, like wars or federal funding cuts). Thick rhetoric is simply pluralistic sociology of knowledge with one eye open for what does and does not persuade. It analyzes what the discipline is, the nature of its discourse, by using a wide range of theoretical "helpers" including sociology, psychology, political science, hermeneutics, classical rhetoric, and so on. This kind of thick rhetoric is not only consistent with certain aspects of Popperian philosophy, but, as I will argue in section e', it can actually be enriched by it.

In the second (thin) type of rhetoric, not only does persuasion matter in scientific discourse, but "that's absolutely all there is, folks!" All we have is conversation: the only constraints we have on inquiry are conversational constraints, and there is no place to stand outside of conversation in order to judge conversation. This second version of rhetoric is basically a variety of radical postmodernism. I call it "thin" because it is very thin on what the advocates of the rhetorical viewpoint seem to want rhetoric awareness to do: to change the way we look at (and possibly practice) science or economics. As Stanley Fish (1988) argues, if one accepts this all-leveling rhetorical point, then absolutely "nothing will follow" from it. It does not work like Freudian analysis; mere awareness does not liberate. To believe—as both Arjo Klamer and Donald McCloskey seem to believe—that recognition of the rhetorical nature of economics will somehow lead us to change our rhetorical practice, or at least to help us protect ourselves from it, is to miss the point.

> That is the most common mistake made by everyone who has ever been enamored of the rhetorical or deconstructive or neopragmatic line: to think that because we now know that we are in a situation, imbedded, constituted socially, we can use that knowledge to escape the implications of what we now know. I call that error *antifoundationalist theory hope*. (Fish, 1988, p. 27, italics added)

Antifoundationalist theory hope is "the mistake of making antifoundationalism into a foundation, of thinking of it as a lever with which we can pry ourselves away from the world delivered to us by our beliefs" (Fish, 1988, p. 30). Correcting this mistake, abandoning the misplaced hope, leaves us with a very thin rhetoric indeed. In fact, it leaves us with nothing to say. Thus we have two views of rhetoric: a thick view that really amounts to a pluralist sociology of knowledge, and a thin postmodernist view from which nothing follows.

Before returning to the sociology of knowledge explicitly in category e', I would like to make one additional point regarding Popperian

CR and the rhetoric of economics (particularly the view of Donald McCloskey). McCloskey claims some philosophical allegiance to Rorty, but he also continues to be a Chicago school economist when it comes to social explanation and human action.[37] Bartley, a Popperian, also claims allegiance to the Chicago school of economics—at one point, Bartley even refers to "My hero Ronald H. Coase" (1990, p. 106, n. 27)—and Bartley's allegiance to the Chicago school of economics actually seems to be stronger than McCloskey's. For Bartley the Chicago view of rationality, incentives, and markets is clearly integrated into his view of the growth of knowledge and the institutional structure of science. "The central concern of that branch of philosophy known as epistemology or the theory of knowledge should be the growth of knowledge. This means that the theory of knowledge is a branch of economics" (Bartley, 1990, p. 89). For Bartley, epistemology is basically a branch of economics, although only of a certain type of economics: the economics of Adam Smith's *Wealth of Nations* and the modern economics of the Chicago, public choice, and transactions cost schools.[38]

This issue of McCloskey versus Bartley requires much more attention than it can be given here, but it should be noted that more is at stake than simply a concern over being "more Chicago than thou." The question is really the question of the proper role of economics as a theory of agent behavior—including scientist agents—in contributing to our understanding of scientific activity. The sociology of knowledge frequently presupposes individual sociological theories in its investigation of science, and certain psychological theories are often presupposed in writing the biographies of individual scientists. The question that Bartley raises is the question of a similar role for microeconomic theory. Can economics be presupposed in metascience discourse—or the case studies designed to replace metascience discourse—in the way that other social theories are frequently presupposed in the examination of science?[39] As I say, this question requires further study, but it is clearly one of the questions raised by recent developments within the Popperian tradition.

e'. In category e, I introduced two separate schools of thought within the sociology of scientific knowledge: the strong program and the constructivist school. While the separation of these two schools may be rather artificial, there are in fact differences, and I will continue to maintain the distinction. The questions addressed in this section concern the general evaluation of these two schools, particularly as they pertain to economics, and how the Popperian tradition might respond to this work.

One obvious Popperian response to the sociology of knowledge (both schools) would be Popper's own response to the earlier sociology of knowledge literature (that associated with Karl Mannheim) almost a half century ago.

> What the "sociology of knowledge" overlooks is just the sociology of knowledge—the social or public character of science. It overlooks the fact that it is the public character of science and of its institutions which imposes a mental discipline upon the individual scientist, and which preserves the objectivity of science and its tradition of critically discussing new ideas. (Popper, 1961, p. 155)[40]

Popper's point—clearly the precursor to Bartley's position that epistemology should be a branch of economics—is that pristinely objective behavior by the individual scientist is not required in order to have the rational growth of scientific knowledge. It is not from the objectivity of the individual scientists that we should expect scientific knowledge—just as, in Adam Smith's words, "it is not from the benevolence of the butcher, the brewer, or the baker, that we expect our dinner" (1776, p. 14)—but rather it is from the open and competitive critical environment of the scientific community in which that knowledge is produced. "Scientific method itself has social aspects. Science and more especially scientific progress, are the results not of isolated efforts but of *free competition of thought*" (Popper, 1961, pp. 154-55). According to this view (and unlike the view in the sociology of knowledge literature), "socially produced" does not imply "cognitively suspect"; in fact, it is precisely the social nature of its production that guarantees the objectivity of science.

While both Popper and Bartley would defend such an argument that the social nature of scientific inquiry need not bias (as the strong program argues) or constitute (as the constructivist school would say) the theoretical products of that inquiry, I *do not find this to be the best Popperian response* to the sociology of knowledge in the current context. The reason is that the differences here—Popper and Bartley on one side, and the views of most sociologists of knowledge on the other—are really differences about *social science*. If one is a microeconomist or one is generally sympathetic to the microeconomic approach to explaining social phenomena—in this case, scientific theories—then it seems quite natural to see the phenomena as a socially optimal result emerging from the competitive process with individually rational agents. On the other hand, if one has Marxist theoretical preferences, then one is not going to see social optimality emerging from self-interested behavior, but rather the reproduction of the existing structure of power and social domination through control of the knowledge production process. And finally, if one is an anthropologist preferring the participant-observer method, then the entire discussion of optimality is simply a category mistake. One can only immerse oneself into the particular scientific culture in an attempt to understand the deeply contextualized social matrix that constitutes the scientific practice. The point is that these are simply three different ways of *doing social science;* and in the process of our choosing any one of them, the decision about how we explain/understand social phenomena *has already been made.* Approaching questions about the methodology of social science by way of a social analysis of natural science practice that presupposes a particular methodology of social science clearly seems to beg far more questions than it answers.[41]

In an effort to avoid this potential quagmire of questions, it is better to "respond" to the sociology of knowledge in a more indirect way.[42] First, it will be argued that the sociology of knowledge exhibits a type of (potentially vicious) circularity that may not be present in the Popperian approach.[43] Second, Bartley's Popperian approach can be compared to the sociology of knowledge when the domain of inquiry is the practice of economics. In this latter case it is possible to argue that Bartley's view actually subsumes much of the sociology of knowledge; that is, the Popperian approach accommodates the type of analysis that occurs within the sociology of knowledge, but it also accommodates analysis along another more normative dimension as well.

The circularity problem arises within the sociology of knowledge because, in Alexander Rosenberg's apt phrase, "this sort of sociology pulls itself down by its own boot straps" (1985, p. 380). Sociologists of knowledge argue that science is nothing more than

> a social institution, and must be understood as such. But if this argument is correct, it must be self-refuting. If scientific conclusions are always and everywhere determined by social forces, and not by rational considerations, then this conclusion applies to the findings of the sociologist of science as well. (Rosenberg, 1985, p. 379)[44]

It seems a bit ironic that the sociology of knowledge—the program that is often touted as "debunking" our positivist notions of science—is hoisted on its own petard in much the same way that positivism hoisted itself. Positivists argued that any claim that is not empirically verifiable is meaningless; but this claim about what is meaningless is itself not empirically verifiable, and therefore meaningless. Similarly for the sociology of knowledge: all scientific claims are socially constituted and not epistemically privileged, but of course this claim itself comes from a community of social scientists and is therefore not epistemically privileged. While this particular circularity may not be as vicious as it first appears, it is nevertheless a persistent problem for the sociology of knowledge.[45]

For the second point regarding the application of the two views to economics, it is useful to recall Bartley's characterization of the research program generated by his approach.

> The new problem of rationality—of criticism and the growth of knowledge—now becomes the problem of the *ecology of rationality*. Instead of positing authorities to guarantee and criticize actions and opinions, the aim becomes to construct a philosophical program to foster the growth of knowledge and to counteract intellectual error. Within such a program, the traditional "How do you know?" question does not arise. For we do not know. A different question becomes paramount: "How can our lives and institutions be arranged so as to expose our positions, actions, beliefs, aims, conjectures, decisions, standards, frameworks, ways of life, policies, traditional practices, and such like, to optimum examination, in order to counteract and eliminate as much error as possible?" (Bartley, 1990, p. 240)

Now suppose this approach is applied to the economics profession. Within this general framework, one could consider any of the sociological variables—class, power, gender, career goals, status, and so on—that one might consider in the sociology of knowledge. In addition, one could consider the influence of ideology, religious beliefs, ethical beliefs, or "physics envy" (Mirowski, 1989) within this general framework. But while one could include any of these broadly sociological factors in the analysis of what goes on in the economics profession, one could also, on the lines of the Bartley approach, *criticize that activity from the viewpoint of knowledge*. One could do the sociology of knowledge or what was called "thick rhetoric" in category d' and still have a vehicle for critically appraising the activity within the discipline. In this view, however, one does *not have methodological rules* that can be rigidly applied to science. What does and what does not contribute to a healthy critical environment is a very textured and context-specific question requiring careful study in each individual case. This view is *not* the meat axe of the demarcation criterion and it is not what McCloskey calls "3' x 5' card philosophy of science" (1989b), but it is also not a view that surrenders entirely the notion of knowledge.

How might the practice of economics be criticized from this view? We already have some examples: Caldwell (1982, 1988a, and 1991a) and Redman (1991). Both Bruce Caldwell and Deborah Redman argue for increased theoretical pluralism within economics from a specifically Bartleyan perspective.[46] Caldwell suggests that the mainstream confront alternative traditions, such as Austrian and institutional economics; while Redman emphasizes social groups such as women, minorities, and foreigners. In both cases, increased pluralism and diversity is advocated—but advocated on the basis of increased criticism and thus the ultimate growth of economic knowledge. This work redefines the methodological goal as *"the provision within economics of an environment in which the optimal amount of criticism is able to flourish"* (Caldwell, 1991a, p. 28). Such methodological appraisals have none of the arrogance associated with strict methodological rules, while at the same time the epistemic link is not totally severed as it is in much of the sociology of knowledge or thin rhetoric.

None of the arguments made above in categories a' through e' require the Popperian tradition to be totally without problems. Clearly, as will be suggested in the conclusion to this chapter, there remain a great number of potential difficulties within the Popperian tradition. The argument being made here is simply that the Popperian tradition is capable of addressing many of the issues raised by recent antimodernist discourse, and it actually *addresses* them rather than taking the ostrichlike posture that is so popular in much of the philosophy of science. Moreover, it does so without giving up on the concept of truth, the role of severe empirical testing, the idea that knowledge grows, or the ability to appraise particular research strategies. The Popperian tradition also does this in a way that not only pertains to, but actively involves, economics.

The next section will also argue in favor of the Popperian tradition as an overarching philosophical framework, but in a much more restricted context: the context of the debate over the role of commonsense notions in social science. This discussion will focus specifically on microeconomics, and the frame of reference will be restricted so as to avoid the more global antimodernist issues that have been considered in this section.

ELIMINATIVE NATURALISM IN THE PHILOSOPHY OF SOCIAL SCIENCE

Consider our commonsense/everyday way of explaining human behavior. I am writing this in my office on a Sunday evening. How do I explain my action of coming into my office on such a day and at such a time? The explanation is quite simple: I came into the office in order to start writing the third section of Chapter 11. I had a certain *desire*—

the desire to complete Chapter 11 and thus be able to send the finished manuscript off to the publisher—and I had certain *beliefs:* in particular, the belief that if I came into the office on Sunday evening I would in fact be able to start writing the third section of Chapter 11. My action is explained by these desires and beliefs; I had a certain desire, a certain goal, and I believed that the action taken would help me satisfy that desire, achieve that goal. The explanation for my returning to the office tomorrow morning will be of the same *form*—it will involve beliefs and desires—but the beliefs and desires will be different. My desire tomorrow morning will be to teach my class, and my beliefs will include the belief that appearing in the same room at the same time that my students are assembled is a reasonable way of achieving my desired goal. Such *explanations* are called *folk psychology* and they constitute the vast majority of the explanations we provide in our everyday lives: explanations of our own behavior, that of our family members, of those whom we meet on the street, of our ancestors, of those in the media, of other professionals, and of people in general. Explanations based on folk psychology involve purposefulness or intentionality; they say, quite simply, that "people do the things they do roughly because they want certain ends and believe these acts will help attain them" (Rosenberg, 1988a, p. 15).

Of course there are many other types of behavioral explanations that we could provide, and do provide from time to time. Many of us would say that our actions must be connected to underlying physical processes in our bodies and brains; sometimes we even use these processes to explain our behavior—for instance, when we are tired or sick. Most of us also recognize that we are the product of an evolutionary process and that certain aspects of our behavior—for instance, those involving food or sex—might be explained on the basis of our genetic coding. Maybe some of our behavior is the result of conditioning by our parents or teachers. (In my own case, this probably explains my aversion to certain sports.) Maybe our behavior is explained by an addiction to a particular chemical or drug such as alcohol, nicotine, or caffeine. Maybe our behavior is determined by our astrological sign— perhaps my desire to get my finished manuscript to my editor on time is merely the typical Taurus response to deadlines. As I say, explanations of human behavior can take many forms.

Despite all of these and many other more or less persuasive alternatives, most of our everyday explanations of human behavior are of the folk psychological type. Not only are such explanations quite commonplace in contemporary society, but they appear to have been equally popular throughout our civilized history. Our commonsense explanations of why people do what they do are substantially the same as those offered by the Greeks. The difference is that, since the Greeks lived in a world punctuated by the intervention of jealous and conspiring gods, their use of desire and belief often extended to the world of nature as well. In addition to being with us for a long time, these concepts from folk psychology also play an essential role in our own sense of self and how we perceive the sensibility or intelligibility of the world around us. Finally, it is important to note that these concepts are also deeply embedded in our moral and judicial codes. After all, the only difference between a tragic but innocent hunting accident and a cold-blooded murder is the degree of intentionality involved.

This intentional influence does not stop at the gates of science. Unlike in physics and biology where "folk" notions often prevail in the mind of the public but entirely different theoretical constructs are employed by practicing scientists,[47] much of social science merely elaborates on, tries to unpack, and attempts to render lawlike the basic notions of folk psychology.

> [In the] social sciences there has been almost universal agreement that the descriptive categories that common sense has used since the dawn of history are the right ones. Traditionally, what we have wanted to know in social science is the causes and consequences of our *actions*, and we hold that these actions are determined by our *desires* and our *beliefs*. Accordingly, social scientists have long searched for laws connecting actions, beliefs, and desires, on the venerable conviction that these are the natural categories into which human behavior and its causes fall. (Rosenberg, 1988a, pp. 11-12)

This tendency to utilize the basic categories of folk psychology is not restricted to sociological studies (e.g., Max Weber's use of Calvinist beliefs to explain the rise of capitalism, or Emile Durkheim's use of the society's desire for social integration to explain the variation in suicide rates); *it is a fundamental component of the standard microeconomic approach to human behavior.* In economics, it is normal to assume that individuals have desires and beliefs. Desires include preferences and the goal of obtaining the most preferred bundle of goods; beliefs include knowledge of all the relevant constraints the individual faces, as well as the belief that consuming the most preferred bundle will actually fulfill the individual's desires. The maximization of an objective function subject to constraints—which characterizes much of microeconomics—is simply a special case of the more general folk psychology of attempting to achieve desired goals subject to the beliefs the agent holds. Again quoting Rosenberg, "rational choice is just folk psychology formalized" (1988a, p. 65).

The point about microeconomics becomes particularly clear when folk psychology is given a more structured characterization. Such a characterization is provided by Rosenberg's [L] scheme.

[L]   "Given any person x, if x wants d and x believes that a is a means to attain d, under the circumstances, then x does a" (Rosenberg, 1988a, p. 25).

While individual details and a number of specific restrictions would need to be added to [L] for it to be applied in any particular case, [L] adequately captures the spirit of an explanation in terms of beliefs and desires. Since the beliefs and desires involved in the explanans are generally considered mental states, if such explanations are to be *causal*, then the kind of causality involved would seem to be *intentional causality*.[48]

If we compare [L] with the schematic presentation of Popper's method of situational analysis (SA) given in Chapter 8, we find that Popper's scheme is simply a more structured version of [L]. In Popper's SA, the rationality principle (RP) that serves as the general law connecting the agent's situation with his or her action is stated explicitly. In [L], the fact that the agent's "situation" involves both beliefs and desires is explicitly stated, but the "law" that the individual will actually do what he or she believes will achieve the goal is only implicit. As argued in Chapter 8, it is only a short step from Popper's SA to the standard microeconomic explanation of individual behavior. The step simply involves a rather restrictive characterization of the *type* of beliefs and desires that can be included in the agent's situation.[49] Thus while microeconomic explanations of individual behavior are much more restrictive than the general form [L], it should be apparent that they are merely special cases of this general explanatory schema.[50]

On the face of it, one might see this intimate relationship between folk psychology and microeconomics to be beneficial to both sides. Microeconomics, with its aggressive mathematical formalism, its powerful econometrics, and its large share of the Nobel prizes in economics, might help bolster the scientific status of folk psychology and intentional explanations more generally against the attacks of their perennial critics such as the behaviorists.[51] On the other hand, microeconomics, long accused of being narrowly hedonistic and rapaciously utilitarian by many of those in other social sciences or the humanities, could point to its humble and sensitive origins in folk psychology. After all, how could microeconomics be so dangerous if it is just a slight formalization of three-thousand-year-old common sense? It certainly seems that folk psychology could gain by looking more scientific, while microeconomics could gain by looking more hermeneutic—a potential Pareto-improving reallocation of disciplinary metaphors. Unfortunately, this symbiosis is not the present relationship between folk psychology and microeconomics.

Folk psychology is currently the target of a sustained and rather harsh critique. This criticism has drawn inspiration from a number of different sources. Some has originated from work in artificial intelligence, while other criticism has come from recent developments in neurophysiology. Some critics simply make the empirical argument that folk psychology has not shown anything like the predictive or explanatory progress exhibited by the natural sciences. There are also a number of different philosophical traditions represented within this critical literature. Some critics are inspired by Quine's naturalized epistemology, and others (sometimes the same individuals) are inspired by traditional materialism.[52] Many simply take the position that, while it is now clear that the solutions offered by behaviorist psychology were inadequate, behaviorism's diagnosis of the problem was basically correct. Although this critical literature has become rather extensive,[53] economics is almost never mentioned—or perhaps more importantly, indicted—in these critiques. The one exception is the work of Alexander Rosenberg (1981, 1983, and 1988a), where the central theme has been that since economics is clearly the best effort to make folk psychology into a scientific theory of human behavior, and since economics has not demonstrated any progress or predictive improvement over the past one hundred years, maybe it is time to abandon folk psychology as the foundation for a scientific theory of human behavior.[54] Let me quickly summarize some of the specific criticisms that have been offered by Rosenberg and others.

The first problem is a problem of *regress*. In order to use an explanatory scheme like [L], it is necessary to know what people believe. The easiest way to find out what people believe is to ask them, but the speech act of individuals is subject to the same scheme [L], and whether they tell the truth about their beliefs or even know the truth about their beliefs will depend on other beliefs that they hold. Thus, as Rosenberg argues, in order to "identify these beliefs and desires, we need to make assumptions about other beliefs and desires. But our original problem was that of determining exactly what people believe and want. If to do this, we already need to know many of their other desires and beliefs, then our original problem faces us all over again" (1988a, p. 33).

The second problem is the problem of *testing*, that is, testing either the initial conditions or the "laws" involved in [L].

If we know what someone's beliefs and desires are, then [L] will tell us what actions he will undertake. If we know what actions the person performs and we know his beliefs, then [L] will tell us what his wants are. And if we know his wants and what actions he performs, then [L] will tell us what he believes. But without at least two of the three, belief, desire, and action, the third is not determinable. (Rosenberg, 1988a, p. 34)

We have encountered this problem before in earlier chapters where SA and the RP were discussed. The rationality principle is either false or unfalsifiable; in either case, it is ill suited to serve as the animating general law in an SA explanation. SA is simply a more structured version of [L] in which the law, the initial conditions, and the deductive structure are more explicit; [L] implicitly harbors all of the testing problems associated with the RP. Mentioning SA in this context provides a hint regarding the way that folk psychology will be linked up to the Popperian tradition later in this section, but for now let us continue to neglect Popper and focus explicitly on the recent criticisms of folk psychology.

The third argument against folk psychology is what I would call the *critique from genealogy*. At one point in Western history, all we had were folk theories; and scientific knowledge has, over the past three thousand years, steadily eroded their domain. It would be a miracle, the argument goes, if our ancient ancestors had somehow hit on a correct theory of human behavior when their folk wisdom was so wrong about everything else. This point of view is exemplified by the following quote from Stephen Stich.

Folk psychology really is a *folk* theory, a cultural inheritance whose origin and evolution are largely lost in prehistory. The very fact that it is a folk theory should make us suspicious. For in just about every other domain one can think of, the ancient shepherds and camel drivers whose speculations were woven into folk theory have a notoriously bad track record. Folk astronomy was false astronomy and not just in detail. The general conception of the cosmos embedded in the folk wisdom of the West was utterly and thoroughly mistaken. Much the same could be said for folk biology, folk chemistry, and folk physics. However wonderful and imaginative folk theorizing and speculation has been, it has turned out to be screamingly false in every domain where we now have reasonably sophisticated science. Nor is there any reason to think that ancient camel drivers would have greater insight or better luck when the subject at hand was the structure of their own minds rather than the structure of matter or of the cosmos. (Stich, 1983, pp. 229-30)

Notice that this genealogical argument does not say that folk psychology is necessarily incorrect; it simply says that discovering it to be so in the future is, based on the lessons of history, a very reasonable possibility. The fourth argument I want to discuss does actually go further; it does argue that folk psychology is *incorrect*—in particular, that the fundamental concept of belief is not a natural kind. Rosenberg makes this argument on the basis of the failure of social science to uncover any reliable lawlike regularities,[55] and Patricia Churchland makes a similar argument from the viewpoint of contemporary neuroscience. As Churchland, referring to her initial work on the topic, writes,

> I argued that consciousness, as it is circumscribed in folk psychology, probably is not a natural kind, in much the same way that "impetus" is not a natural kind. Nor, for example, do the categories "gems" or "dirt" delimit a natural kind. That is to say, *something* is going on all right, but it is doubtful that the generalizations and categories of folk psychology either do justice to that phenomenon or carve Nature at her joints. (Churchland, P. S., 1986, p. 321)

While these four sets of arguments do not exhaust the recent criticisms of folk psychology, they do represent a good sample of the general class of problems that have been discussed. And there have been a number of different responses to these criticisms. They range, on one extreme, to the outright defense of intentional notions and their importance in any reasonable explanation of human behavior, to, on the other extreme, eliminative materialism and the complete abandonment of our mentalistic vocabulary of belief and desire.[56] I would particularly like to focus on the most radical of these responses: *eliminative materialism*. This view—according to Patricia Churchland, one of its strongest advocates—consists of three separate theses:

1. "that folk psychology is a theory";

2. "that it is a theory whose inadequacies entail that it must eventually be substantially revised or replaced outright (hence 'eliminative')"; and

3. "that what will ultimately replace folk psychology will be the conceptual framework of a matured neuroscience (hence 'materialism')" (Churchland, P. S., 1986, p. 396).

The first thesis is necessary because, if folk psychology is not a *theory* of human behavior, then there is nothing to replace or eliminate. Poetry and fiction are ways of talking about humans, but no one would argue that they should be replaced by "matured neuroscience" since (among other things) they were never offered as a *theory* of behavior. This creates some confusion in the literature because it puts eliminative materialists—the group most hostile to folk psychology—in the rather awkward position of defending folk psychology as a *theory* of human behavior. The second thesis is much like the behaviorist critique of folk psychology. The eliminative materialist solution of thesis 3 (and the way it interacts with thesis 2) is different from that of behaviorist psychology, but the critique in thesis 2, taken alone, is much the same. The interaction of thesis 2 and thesis 3 is quite clear in the following characterization of eliminative materialism by Paul M. Churchland.

> On this view, folk psychology is not just an incomplete representation of our inner natures; it is an outright *mis*representation of our inner states and activities. Consequently, we can not expect a truly adequate neuroscientific account of our inner lives to provide theoretical categories that match up nicely with the categories of our common sense framework. Accordingly, we must expect that the older framework will simply be eliminated, rather than be reduced, by a matured neuro-science. (Churchland, P. M., 1984, p. 43)

In this view, then, folk psychology and thus any social science based on its categories is simply out. Maybe we will need to hold onto some of these folk notions for a while; but in time, and perhaps quite soon, they will be eliminated.

The eliminative materialist view of the future of folk psychology is mirrored in Alexander Rosenberg's view (1981) of the future of economics and most of social science.[57] In Rosenberg (1981) it was argued that the concepts of belief and desire should be eliminated from social science (economics specifically), and that they would be replaced with the more scientific concepts from biology—particularly sociobiology. Rosenberg is not strictly in the eliminative materialist camp since his inclusion of biology would require the modification of thesis 3 above, but his argument is clearly in the same eliminative naturalist spirit as this view. Since it seems unlikely that anyone would argue for the replacement of all of our economic theories by neurophysiology, and since Rosenberg's view does indict economics and is based on the same basic arguments as eliminative materialism, in the following discussion I will refer to it as the eliminative naturalist position. By *eliminative naturalism* I simply mean the view characterized by Patricia Churchland's three theses with an "or another branch of matured natural science" added to thesis 3. Eliminative naturalism argues that our commonsense folk psychology notions are inadequate and that they should be replaced by concepts from natural science. This criticism applies not only to folk psychology per se, but to any explanation involving beliefs and desires, including Popper's SA and microeconomic explanations of the behavior of individual economic agents.[58]

With this rather elaborate groundwork now completed, let me return to the main theme of the Popperian tradition in economic methodology. How do these issues relate to the Popperian tradition? And what would be the best Popperian response to eliminative naturalism?

In addressing this latter question, I would like to start off down what would appear to be the most obvious road, even though we will quickly discover that it is not actually the best road if one desires a Popperian response to eliminative naturalism. This obvious road is *Popper's own writings on the philosophy of mind*.[59] When Popper discusses issues such as the relationship connecting the physical world, the mind, and objective knowledge, he frames the discussion in terms of his "three worlds" ontology.

> The world consists of at least three ontologically distinct sub-worlds; or, as I shall say, there are three worlds: the first is the physical world or the world of physical states; the second is the mental world or the world of mental states; and the third is the world of intelligibles, or *ideas in the objective sense;* it is the world of possible objects of thought: the world of theories in themselves, and their logical relations; of arguments in themselves; and of problem situations in themselves. (Popper, 1972, p. 154)

In this view the third world, the world of theories and objective knowledge, interacts with the physical world—for instance, we use physical theories to build a bridge that exists in the physical world—but it only interacts through the second world, the world of individual thoughts. There is a link from World 3 to World 1, but it is an indirect link through the second world of individual minds. World 3 is thus semiautonomous and cannot be reduced to World 1. Popper extends this argument to World 2: since World 2 interacts with World 3, Popper argues that World 2 cannot be reduced to World 1 either.

> These facts seem to me to establish the impossibility of any reduction of the human World 2, the world of human consciousness, to the human World 1, that is, essentially, to brain physiology. For World 3, at least in part, is autonomous and independent of the other two worlds. If the autonomous part of World 3 can interact with World 2, then World 2, or so it seems to me, can not be reducible to World 1. (Popper, 1982, p. 160)[60]

For Popper this argument implies that "physicalism" or "physical reductionism" (and thus eliminative materialism) views that deny "the existence of Worlds 2 and 3" are a "mistake" and reduce to "absurdity" (1982, p. 161).

While this argument obviously constitutes *a* Popperian response to eliminative naturalism, there are a number of reasons why I do not think it is a very good response. First, the three worlds ontology is one of the shakiest aspects of the entire Popperian philosophical program. I do not have a detailed critique here, but just the simple advice that good fortifications should be built on solid ground. Second, even if one accepts the basic three worlds ontology, there seem to be a number of slips in the logic of the argument, from the partial autonomy of World 3 to the "irreducibility" (to use Popper's term) of World 2 to World 1. Perhaps this argument could be improved so that it is more convincing, but at this point it remains unpersuasive. Third and finally, Patricia Churchland has provided a rather simple reductio ad absurdum of Popper's position. She argues that, if one accepts Popper's position, then not only do mental processes end up nonphysical, but—much less desirably—"mass, temperature, charge, and paradigmatically physical magnitudes generally must also be nonphysical properties" (Churchland, P. S., 1986, p. 342).

Before presenting what I believe is a more adequate Popperian response to eliminative naturalism, I would like to reframe slightly the question of the relationship between the eliminative naturalist critique of folk psychology and the Popperian tradition. Obviously, Popper's SA is indicted by eliminative naturalism, but the real issues run much deeper than merely this. In a sense, the tension between Popper$_s$—the advocate of SA in social science—and Popper$_n$—the falsificationist philosopher of natural science—discussed in earlier chapters (particularly Chapter 6) represents a *Popperian microcosm of the tension between folk psychology and eliminative naturalism.* Popper has a rather strict notion about what constitutes a valid scientific explanation: it is essentially a D-N covering law explanation involving falsifiable scientific laws.[61] And this is a view of scientific explanation that is shared by most of the advocates of eliminative naturalism. Eliminative naturalism basically argues that explanations involving belief and desire do not satisfy the conditions for a valid scientific explanation. Many of the reasons for this, such as the unfalsifiability of the "laws" involved, are exactly the problems discussed in Chapter 6 and elsewhere in this book as sources of tension between Popper$_s$ and Popper$_n$. Some of the issues raised by eliminative naturalism actually go beyond the tension between Popper$_s$ and Popper$_n$—in particular, the materialist attack on substance dualism. But these issues can be separated from the more basic thesis that "folk psychology fails to explain." In what follows, then, I would like to repackage the eliminative naturalist challenge to folk psychology explanations as essentially the same set of issues raised by the tension between Popper$_s$ and Popper$_n$. In this way, the Popperian response to eliminative naturalism becomes simply a matter of finding a way to release the tension between Popper$_s$ and Popper$_n$.

As Bruce Caldwell (1991a) has recently argued, *critical rationalism (CR) provides just such a release of tension between Popper$_n$ and Popper$_s$.*[62] Recall that, according to CR, "scientific theories are distinguished from myths merely in being criticizable, and in being open to modification in light of criticism" (Popper, 1983, p. 7). In this view the problems with the testability of the "laws" as well as the difficulties associated with obtaining information about beliefs can both be overlooked as long as the explanations provided are open to criticism and can be improved on the basis of that criticism. By deemphasizing falsificationism and empirical testing and emphasizing instead a general openness to criticism in a variety of forms, the rigid distinction between a "good scientific explanation" and a "bad scientific explanation" is blurred, and so too is the distinction between Popper$_s$ and Popper$_n$. As Caldwell summarizes his argument,

> Rather than choose between falsificationism and situational analysis, I proposed that Popper's writings on critical rationalism permit an escape from the dilemma. If one is a critical rationalist, one is less interested in such questions as how to demarcate science from nonscience or how to justify one theory as better than another. Instead, the emphasis is on criticism. The type of criticism that one should employ cannot be specified prior to the statement of a problem to be solved; criticism is always problem specific. Within the social sciences, it turns out that the decision to retain the rationality principle is often a very effective way to develop and criticize theories. (Caldwell, 1991a, p. 28)

Of course, this CR solution requires that our explanations in folk psychology or in microeconomics actually get criticized, and actually get revised in light of that criticism. This is clearly the case in our commonsense intentional explanations in everyday life: my desire to teach my class is an inadequate explanation for my coming into the office on Tuesday, once it is pointed out that I do not have a class on Tuesday. While our response to such criticisms are usually to provide an alternative intentional explanation in terms of different beliefs and desires, it is nonetheless a revision in light of criticism. With respect to the microeconomic explanations of individual behavior, perhaps the economics profession has not been as willing to revise its explanations in the light of criticism as CR would recommend, but that simply carries us back to the issue that ended the preceding section: using the CR approach to criticize disciplinary practice in order to enhance the critical environment. This criticism would apply to revision of the model not only within the basic SA mode of explanation, but outside that mode as well. "A critical rationalist would also encourage alternative approaches, those that try to improve on the rationality assumption or that attempt to explain social phenomena without recourse to individual maximizing behavior" (Caldwell, 1991a, p. 28).

Even if an advocate of eliminative naturalism were to accept these arguments (on how CR weakens the impact of criticism about the unfalsifiability of the "laws" of folk psychology and about the possible regress involved in identifying beliefs), they could still make the more fundamental *materialist* argument against folk psychology. If intentional explanations are causal explanations, then there should be a reductive relationship between the mentalistic language of belief and desire and the physicalist language of neuroscience. Since this reductive relationship is not available—and it appears it never will be available—we can conclude that these intentional "kinds" are not natural kinds and that belief and desire simply do not cut nature "at the joints."[63]

One response to such an argument would be simply to note that, while a Popperian CR framework does not necessarily presuppose materialism, it is not at odds with it either.[64] For example, if one characterizes the relationship between the mental and the physical (the mind and the brain) as one of *supervenience* rather than reduction, then one can maintain a commitment to a materialist ontology without requiring the reduction of the mental kinds to physical kinds. In such a framework, the physical makes demands on the mental, but the demands do not require reduction. The tight linkage between the physical and the mental is broken without abandoning the commitment to materialism. Such a view can allow for intentional explanations, intentional causality, and intentional kinds, while still allowing for the primacy of the physical. This characterization of action explanations, which might be called "supervenient folk psychology," is defended at length by Kathleen Lennon (1990). She argues that such a framework "gives human sciences a place which is both scientific and autonomous without at any stage denying the truth of materialism" (1990, p. 14). While such a position is not necessarily Popperian, I do not see why it should be inconsistent with a general CR approach to knowledge.[65]

In conclusion, then, it has been argued that, by reconciling Popper$_s$ and Popper$_n$, CR provides a response to the eliminative naturalist critique of folk psychology, and thus to intentional explanations in microeconomics. While not all of the criticisms of folk psychology are answered in this way, many are—specifically, those involving testing and regress. Other issues, such as accommodating an ontological commitment to materialism, could be added in such a way that they do not produce a direct conflict with the major tenets of the Popperian program. Intentionality remains in the discourse of social science, but individual intentional explanations are continuously subject to critical interrogation and revision. Alternative nonintentional explanatory schemes are also encouraged so that criticism develops between various explanatory schemes as well as between individual explanations within a particular scheme. In this Popperian view, neither philosophical fiat nor recent developments in neurophysiology have the right to mandate the particular form that explanations in the social sciences may take. "Under critical rationalism, the goal is to keep the critical process going, to build an ecology of critical inquiry, an environment in which the optimal amount of criticism is able to flourish" (Caldwell, 1991a, p. 25).

CONCLUSION

I chose to address antimodernism and eliminative naturalism not because they were easy targets for the Popperian tradition, but rather because these views in fact represent, I believe, the views that most need to be addressed by any overarching philosophical framework that purports to provide a backdrop for contemporary discourse in the social sciences. Both of these views represent powerful and radical challenges to the mainstream of modern philosophical and social thought.

Antimodernism comes in a variety of manifestations, from the innocently suggestive to the rapaciously deconstructive. According to some of its more aggressive authors, late-twentieth-century intellectual life is poised at a cusp, on the verge of a fundamental transformation in our way of thinking and talking, the likes of which has not occurred since the Enlightenment. It is argued that radical perspectivism, and truth characterized as Nietzsche's "mobile army of metaphors" (Rorty, 1989b, p. 17), allow us to transcend the Enlightenment and in a sense complete it, by finally abandoning the deity surrogates of universal truth and universal good.

> Once upon a time we felt the need to worship something which lay beyond the visible world. Beginning in the seventeenth century we tried to substitute a love of truth for a love of God, treating the world described by science as a quasidivinity. Beginning at the end of the eighteenth century we tried to substitute a love of ourselves for a love of scientific truth, a worship of our own deep spiritual or poetic nature, treated as one more quasidivinity. . . . The line of thought common to Blumenberg, Nietzsche, Freud, and Davidson suggests

that we try to get to the point where we no longer worship *anything*, where we treat *nothing* as quasidivinity, where we treat *everything*—our language, our conscience, our community—as a product of time and chance. (Rorty, 1989b, p. 22)

Whether this fundamental transformation comes to pass, or not, remains to be seen. In favor of such a transformation, it is clear that antimodernism is a powerful intellectual force, and that it has already fundamentally altered the way we look at and redescribe certain aspects of our social life. Against such a transformation is the argument that we have heard such prophetic rumblings about the coming end of the bourgeois life world many times before. In either case, though, antimodernism is not an intellectual movement that can be neglected by any philosophical position currently campaigning for our attention in the social sciences.

Although the impact of eliminative naturalism has been much more local—restricted to the philosophy of mind and the philosophy of social science—it too represents a rather radical position that would require a fundamental reorientation in the way we view ourselves and our fellow human beings. As long as we have been aware of ourselves, we have couched our actions and those of others in terms of belief and desire. To change that fundamental perception would mean a disruption in our basic world view similar in magnitude to the changes suggested by some antimoderns. "If the empirical presuppositions of folk psychology turn out to be false, as well they might, then we are in for hard times. Deprived of its empirical underpinnings, our age old conception of the universe within will crumble just as certainly as the venerable conception of the external universe crumbled during the Renaissance" (Stich, 1983, p. 246). Thus, eliminative naturalism is also a perspective that deserves the attention of any philosophical position that purports to provide a backdrop for our discussions about human and social action.

The preceding two sections have offered what I believe to be *the best available Popperian responses to these two literatures, that is, to antimodernism and eliminative naturalism.* In both cases, the responses are based on the CR interpretation of the Popperian position. It has been argued that, while foundationalism and methodological rules are no longer our goals, the Popperian philosophical tradition can still provide a useful framework for discussing philosophical questions in economics. This response is particularly important in the case of antimodernism, where most traditional philosophies of science have almost nothing to say. It has been argued that Popperian philosophy in CR form can accomplish the following: effectively dodge Rorty's bullet; be consistent with, and actually generate, a type of sociology or thick

rhetoric of science; surrender foundationalism without self-destruct-
ing or giving up on rationality; and accommodate intentional and SA
explanations within a deductive explanatory framework. Moreover, it
can do all this without abandoning the notions of knowledge or rationality, and do so in a way that involves economics and economic
explanation in a fundamental way. Thus the Popperian tradition presents a rather tough challenge to any other philosophical framework
that might be contesting for this same background position, and it surely
suggests that at least certain parts of the Popperian tradition should be
saved.

While all of these positive things can be and have been said in favor
of the Popperian tradition, it must be admitted that a number of troublesome areas and potential difficulties still remain. I would like to
close by briefly considering two of these problem areas. First, there
are questions about CR itself: at best it needs additional work, and at
worst it ends up being a "contentless directive" (Nola, 1987, p. 497).
Second, there is the whole question of the current meaning and future
path of the Popperian tradition.

I have argued above that CR represents the Popperian tradition's
best bet, that it provides a reasonable response to antimodernism and
eliminative naturalism, and that it is generally pregnant with interesting ideas and worthy of serious consideration as a philosophical research program. Despite all of this, there remain a number of problems
with CR. These issues require additional work, and many philosophers
in the Popperian tradition seem reluctant to fill in the details. Popper,
while endorsing CR in his later work, provided details that were only
intermittently coherent at best. I would like to mention three of these
issues; while these are all things that may be worked out in the further
elaboration of the CR program, at this point they remain problems.

First, there is the issue of realism. I have argued in Chapters 8 and
9, and it is now generally accepted in the literature, that Popperian
falsificationism fails as a realist philosophy of science: there simply
is no systematic connection between following a falsificationist methodology and discovering the real causes of that which is being investi-
gated. Bartley's CR seems to be no better in this regard. The hope in
CR is that "evolutionary epistemology" can be used to link up criticism and the theoretical constructs that evolve out of critical discussion with the underlying real causal mechanisms.[66] At this point, though,
the suggestions of Bartley and others in this regard are just that: suggestions. It may be possible through additional work on evolutionary
epistemology to ferret out a connection between criticism and truth
that can stand up where Popper's efforts to establish a relationship

between falsification and truth failed to stand up, but such a connection is not currently available. Perhaps CR should simply drop all pretensions toward realist philosophy of science; in which case, CR is even closer to antimodernism than was suggested above. If that is in fact the implication of CR, then it needs to be explicitly spelled out. While such a solution would remove the realism problem that plagued Popper and would possibly make some antimodernist friends, I suspect that it would also mean the loss of a number of CR's current supporters.

Second, there is the question of Bartley's (and Gerard Radnitzky's) economics (see Radnitzky and Bartley, 1987). According to the argument that epistemology should be a branch of economics, the process of criticism should, through an invisible-hand-type mechanism, produce knowledge in the same way that the competitive market process in the economy produces efficiency or wealth. While the so-called first fundamental theorem of welfare economics—that every competitive equilibrium represents an efficient allocation—is a standard result in microeconomics, it is not at all clear how, or if, the analogy carries over to knowledge. As every sophomore economics student should know, the first fundamental theorem of welfare economics requires perfectly competitive markets (i.e., all traders take prices as parameters), equilibrium prices, no externalities, no economies of scale, no public goods, and a host of other auxiliary assumptions—assumptions that do not seem to hold, at least automatically, in the domain of knowledge and criticism. One response may be, given the Hayekian and Austrian connection in Bartley's work, that it is not the (Walrasian) first fundamental theorem that is at work in the domain of knowledge, but rather the Austrian market process. If so, then one has the additional problem of clarifying what is meant by an Austrian market process in this (or any other) context. Yet another response might be that it is not economic efficiency at all that is at the heart of the analogy, but rather economic growth and development. Maybe it is not the competitive process in either Walrasian or Austrian terms that holds the secret to the growth of knowledge through criticism, but rather an Adam Smith-type incentive argument; again, one has the problem of working out the forces of economic growth and applying them to this particular problem. In any case, the point is that, while it *may be possible* to show how the critical competition of ideas works "like" the invisible hand of competition in economics, this is not in fact something that has been done in the CR literature. Like evolutionary epistemology, it seems to be an interesting suggestion, but at this point it remains a suggestion.

Third, there is the entire question of the meaning of criticism; this is the issue of criticism being a "contentless directive." What could count as criticism besides pointing out internal contradiction, pointing out contradiction with more fundamental or cherished beliefs, or empirical criticism? This question relates to the previous economic issue, since one really wants productive, or efficient, or optimal criticism, and all of these things would depend on how the growth of knowledge process is characterized in more detail. As in the discussion of realism, there is a solution that pushes CR more into the antimodernist camp. It is the position that "criticism" cannot be specified in advance, that what counts as criticism is inexorably context specific and socially constituted.[67] Again, while such arguments can be offered, a lot more work will be required to spell out these ideas in detail and to make certain such a characterization of criticism does not introduce tensions with other aspects of the Popperian tradition.

Finally, there is a more general question about the current standing of the Popperian tradition that is really independent of any of these three potential difficulties with CR. *The problem is that it is simply not clear that the terms* Popperian tradition *or* Popperian program *capture anything particularly cohesive anymore.* Lakatos and Bartley are gone. John Watkins (1984) has proposed a neo-Popperian program where truth plays only a minimal role and the methodological prescriptions are corroborationist, rather than falsificationist. Ernest Gellner (1974, 1985) argues for a type of falsificationism that seems more positivist than Popperian: empiricism is for Gellner not only that which demarcates science from nonscience, but that which demarcates the West from the rest. On the other hand, Elie Zahar (1983, 1989) has abandoned Popper's notion of the empirical basis entirely and leans toward a type of phenomenalism that, when combined with the Zahar-Worrall notion of novel facts, produces a "Popperian" program devoid of both falsificationism and Popper's conventionalism. John Worrall, in a recent rather pessimistic paper (1989b), has argued that both Popper and Watkins have failed to solve *the* problem of the Popperian program: the problem of induction.[68] While CR seems to be the interpretation of the Popperian tradition with the most general support, there are many problems, as discussed above, and also a number of individual versions: Bartley, Radnitzky, Popper, and now Musgrave (1989). Each of these seems to emphasize different aspects of CR and, as the comments on realism and criticism immediately above suggest, there is the possibility of a quite antimodernist interpretation of CR that, while independently interesting, may clash with other aspects of the Popperian program (such as realism). All in all, there seems to be disarray within the tradition.

None of these last few paragraphs should detract from the main argument in this chapter. CR is the heir apparent to the Popperian tradition; and while it is not without problems, it provides a quite reasonable response to both antimodernism and eliminative naturalism. This is not a claim that can be made by any other philosophical position that started its life explicitly as a philosophy of science. The Popperian tradition will continue to have an important role to play in our philosophical discourse about economics. The days where almost everyone preached falsificationism are gone, as are the great novel-fact hunts of the 1980s; but the Popperian tradition will continue to have a role in economic methodology, nonetheless.

NOTES

1. Particularly notes 24, 26, and 31.
2. This general approach, by the way, will be viewed as a modernist response to such complexity by some antimodernist readers.
3. Kuhn, however, denies that his arguments regarding theory-ladenness and incommensurability lead to relativism:

> My critics respond to my views on the subject with charges of irrationality, relativism, and the defense of mob rule. These are all labels which I categorically reject, even when they are used in my defense by Feyerabend. To say that, in matters of theory-choice, the force of logic and observation cannot in principle be compelling is neither to discard logic and observation nor to suggest that there are no good reasons for favoring one theory over another. (Kuhn, 1970c, p. 234)

4. Also see Rorty (1982a, 1988, 1989a, and 1989b) and the essays in Malachowski (1990).
5. The term *post-Philosophy* might have been used rather than postepistemology to characterize this view since Rorty's critique encompasses the aesthetic and ethical domains of Philosophy as well as the epistemological. The reason for focusing on epistemology is that, for Rorty, epistemology (perhaps Epistemology) is the metadiscourse for all of modern (i.e., post-Kantian) philosophy. What Rorty criticizes is "epistemology-centered philosophy" (1979, p. 390) or philosophy-as-epistemology (p. 136), and thus *postepistemology* seems to be an apt characterization.
6. Not only is the critique the part of his argument most readily associated with Rorty's name, it also seems to be less controversial than his suggestions regarding pragmatism and hermeneutics. For example, John D. Caputo (1983), Peter Munz (1987), and Georgia Warnke (1985) all criticize Rorty's interpretation of the hermeneutic tradition, while Alexander Rosenberg (1989) expresses similar concerns about his presentation of pragmatism. Such arguments have prompted Daniel Dennett to posit the "Rorty Factor": "Take whatever

Rorty says about anyone's views and multiply it by .742" (Dennett, 1982, p. 349).

7. Cohen and Dascal (1989) and Baynes, Bohman, and McCarthy (1987), for example.

8. See note 10 below.

9. As "radical postmodernism" is used here, it is difficult to associate with any narrow set of authors, although Jacques Derrida, Michel Foucault, and Jean-Francois Lyotard would possibly qualify. (Rorty is frequently included, but his position has been, and should be, defined more narrowly.) Historically, this version of postmodernism draws its inspiration from the work of Martin Heidegger, Friedrich Nietzsche, and the later Ludwig Wittgenstein.

In a recent book on postmodernism and social science, Pauline Rosenau (1992) subdivides postmodernism into "affirmative" and "skeptical" orientations. Her skeptical category would roughly fit what I call "radical postmodernism," while her affirmative postmodernism might appear in a number of my different categories.

10. Some recent works on economics that explicitly use the term *postmodernism* include Amariglio (1990), Backhouse (1991a), Klamer (1987), Milberg (1988), Mirowski (1991), Rossetti (1990), and Ruccio (1991). More implicitly, some of these postmodernist ideas have influenced recent historical works such as Mirowski (1989) and Weintraub (1991).

11. Work sympathetic to this general approach would include Klamer (1983, 1984, 1988a, 1988b, and 1990), Klamer and McCloskey (1989), and McCloskey (1984, 1985, 1988a, 1988b, 1988c, 1989a, and 1989b); while criticisms include Backhouse (1991a, 1991b), Caldwell and Coats (1984), Coats and Pressman (1987), Davis (1990), Dyer (1988), Hausman (1992), Hollis (1985), Mäki (1988a, 1988b, and 1991b), Rappaport (1988a and 1988b), Rosenberg (1988a and 1988b), and a number of the papers in Klamer, McCloskey, and Solow (1988).

12. This quote uses "relativist" more generally than it was used in category a above in the text.

13. The range becomes even wider if we add related literature such as "social epistemology" (e.g., Fuller, 1988).

14. Neither of these schools of thought really includes the classical sociology of knowledge associated with Karl Mannheim in the 1930s. This classical approach differs substantially from both of the modern schools discussed in the text (see Susser, 1989). In addition, during the past few years, a number of different approaches have appeared in sociology and the history of science that are related to the "schools" considered in the text, but do not fit very comfortably into either one of them (Galison, 1987, for example).

15. Influential works in the strong program include Barnes (1974 and 1977) and Bloor (1976); contributions to the constructivist program include Collins (1985), Knorr-Cetina (1981), and Latour and Woolgar (1986), among others. Pickering (1992) is a recent collection of papers where a number of these different views are represented. There is a small but growing literature discussing the sociology of scientific knowledge in relation to economics; it

includes Backhouse (1992), Burkhardt and Canterbery (1986), Coats (1984, 1988), Collins (1991), Mäki (1992), and Weintraub (1991).

16. In some cases, these microsocial factors end up being what most economists would call "microeconomic factors." For example, in discussing particle physics, Karin Knorr-Cetina defines

contingency in terms of a negative relationship of dependence between two desired goals, or research utilities, such that one utility can only be obtained or optimized at the cost of the other. In this situation, particle physicists resort to a strategy of commerce and exchange: they balance research benefits against each other, and they "sell off" those which they think that, on balance, they may not be able to afford. Particle physicists refer to this commerce with research benefits as "trade-offs." (Knorr-Cetina, 1991, p. 113)

17. David Bloor clearly says that his critics have "failed to see that I am an inductivist" (1984, p. 83).

18. Bloor expands his point in the following way.

The student of the piano may not be able to say what features are unique to the playing of his teacher, but he can certainly attempt to emulate them. In the same way we acquire habits of thought through exposure to current examples of scientific practice and transfer them to other areas. Indeed some thinkers such as Kuhn and Hesse believe that this is exactly how science itself grows. Thought moves inductively from case to case. (Bloor, 1984, p. 83)

19. In some ways the constructivist position is related to the "ethnomethodological" approach of Harold Garfinkel (1967). In fact, constructivist themes overlap with recent ethnomethodology-inspired studies of scientific "work activity" such as Garfinkel, Lynch, and Livingston (1981), Livingston (1986), and Lynch (1985). The relationship between these two approaches to science studies—the sociological and the ethnomethodological—is debated in the exchange between Bloor (1992) and Lynch (1992a and 1992b).

20. In general, it is very difficult to place the sociology of knowledge relative to the categories of relativism, postepistemology, and radical postmodernism; in a sense, it depends on which particular author one is considering. For example, Bruno Latour and Steve Woolgar (1986) sound very close to radical postmodernism despite the claim to the contrary in Latour (1990), while historians of science such as Peter Galison (1987) seem to be responding more to questions raised by Kuhnian relativism.

21. Bartley's main argument is presented in Bartley (1982 and 1984 [1st ed. 1962]), and summarized briefly in Bartley (1987a and 1990, pp. 229-55). He claims that his "approach is based on, interprets, partially corrects, and generalizes Popper's approach" (Bartley, 1987a, p. 211).

22. With respect to Bartley's interpretation, Popper writes,

It was only recently that I began to suspect this, and to suspect that my own approach to the theory of knowledge was more revolutionary, and for that reason more difficult to grasp, than I had thought. This suspicion arose from a new way of viewing my own approach, and its relation to the problem situation in philosophy; a way that was suggested to me by my friend W. W. Bartley III. His views are striking in themselves. But they also explain why certain misunderstandings of my position are almost bound to arise. (Popper, 1983, p. 18)

23. Often what is called "justificationist" by Bartley and by others in the Popperian tradition should more properly be called "foundationalist": the idea that scientific theories are built up from solid (usually empirical) foundations. However, in the interest of consistency with quoted sources, I will continue to use the term in the way that it is used in the Popperian tradition even though it may be at odds with standard philosophical usage.

24. While Bartley may be correct that nonjustificationism reflects the senior Popper, and that it is generally consistent with the Popperian philosophical tradition, I certainly cannot agree with him regarding the filiation of Popper's ideas on these matters.

Bartley argues that, in "some of Popper's early works, there are occasional passages which might lead one to count him as a limited rationalist, or even as a fideist," but "these early fideistic remarks are relatively unimportant; they play no really significant role in Popper's early thought and none at all in his later thought" (1987a, p. 211). This fideism, according to Bartley, was abandoned in 1960 when Popper adopted a nonjustificationist view. The "fideism" that Bartley discusses is roughly equivalent to what was above called "radical postmodernism": it is a view that "is frankly irrationalist"; it "joyfully takes any difficulties to mark the breakdown of an overreaching reason" (1987a, p. 208). Thus, according to Bartley, Popper started out as a type of radical postmodernist, but it didn't matter because by 1960 he had accepted Bartley's more moderate nonjustificationist view.

I have a number of difficulties with Bartley's "Popper as postmodernist," no matter where this tendency is placed in his intellectual development. I would tell the story differently; in fact I *did* tell the story differently, in Chapter 9 of this book. My story is basically a story of three Poppers: a young Popper, who held generally justificationist preferences but recognized the difficulties of connecting up these preferences with his methodological insights, and therefore chose simply to suspend epistemological discourse; a middle-aged Popper, who was more bold regarding justificationism because he felt that verisimilitude would be able to supply the previously absent connection; and a senior Popper, who embraced Bartley's view as a way of holding onto rationality in the face of failed verisimilitude and the relativist/postempiricist/postmodernist onslaught.

25. This distinction between justifying a theory and justifying one's belief in the theory is rather subtle. Allan Musgrave attempts to make the distinction clear:

Talk of justifying or giving reasons for beliefs (acceptances as true, adoptions as true, etc.) is ambiguous between justifying or giving reasons for *what is believed* and justifying or giving reasons *for the believing of it.* Popper's contention is that we cannot justify the things we believe but *can* sometimes justify our believing them; that we can give no good reason for what we accept as true, but can sometimes give a good reason for our accepting it (tentatively of course). Critical rationalism is the view that the failure of our best efforts to show a theory false is a good reason for *us to accept* it tentatively as true, without being any reason at all *for the theory itself.* (Musgrave, 1989, p. 307)

26. While the text provides a generally sympathetic reading of the CR, I would like to note here where I think the argument is the strongest and where it is the weakest. Bartley's CR can be decomposed into three separate propositions. (You could add a fourth that says, "This is really Popper's view.") These three propositions might be characterized as follows:

(i)   Justificationism has traditionally dominated philosophy, and it has failed.
(ii) What is needed is a way of rationally evaluating beliefs and theory choices that is nonjustificationist.
(iii) Criticism is the key to the solution of the problem in proposition (ii).

While Bartley supports all three of these propositions, (iii) seems to be far more difficult to defend than (i) or (ii). Proposition (i) seems to be a simple fact of late-twentieth-century intellectual life; and given proposition (i), proposition (ii) seems to be fundamental to saving the general project of Enlightenment rationality. Proposition (iii) has a certain friendly obviousness, but it is far less firm. Its support depends greatly on the absence of any other palatable solutions to proposition (ii) and on the (as yet inchoate) arguments regarding "evolutionary epistemology" (see note 31 below). For a reasoned criticism of proposition (iii) from within the Popperian tradition, see Gellner (1985, esp. pp. 37-39). Of course, "justification" in propositions (i) and (ii) should perhaps be replaced with "foundationalism" (see note 23 above).

27. For a general discussion of the different concepts of "realism," see Nola (1988); and for a discussion of these different concepts of realism with a particular emphasis on economics, see Mäki (1989a).

28. Popper is quite clear in admitting this:

I know that I am incapable of creating, out of my own imagination, anything as beautiful as the mountains and glaciers of Switzerland, or even as some of the flowers and trees in my own garden. I know that ours is a world I never made. . . . I can only repeat that this argument satisfies me; perhaps because I never really needed it: I do not pretend that I ever doubted the reality of other minds, or of physical bodies. (Popper, 1983, p. 84)

29. Although some defenders of Popper's CR now argue that his correspondence view of truth is "superfluous" (Munz, 1985, p. 246).

30. A clear statement of the conjectural realist position is given in Worrall (1982, pp. 229-30); Worrall presents his own modification of Popper's view, called "structural realism" in Worrall (1989a). The guarded nature of this position led one recent author to characterize conjectural (or critical) realists as "chastened moderns" (Murphy, 1990, p. 296).

31. Recall that, with the failure of verisimilitude, there is no linkage between the methodology that Popper recommends and achieving the cognitive goals of science; Popper and his students have never been able to demonstrate that following his methodology will provide us with truth (even approximately) about the world.

The solution to this problem posed by a number of authors in the Popperian tradition is the "evolutionary epistemology" of Donald Campbell, Popper, and others (see Munz, 1985 and 1987, and the papers in Radnitzky and Bartley, 1987). Briefly stated, the argument is that, since animals with more accurate representations of the world have a survival advantage over those with less accurate representations, survival involves the selection of cognitive structures that more accurately represent. For humans, our theories about the world are part of cognitive structures, and those theoretical structures will be selected in the same way. As Popper himself says, "Starting from scientific realism it is fairly clear that if our actions and reactions were badly adjusted to our environment, we should not survive" (1972, p. 69). Now—the argument goes—the selection mechanism, the method of trial and error, is the same in the domain of knowledge as it is in evolutionary biology: variation and selection, or conjectures and refutations.

> And both, so Popper maintains, are produced by the same Darwinian mechanism: the highest creative thought, just like animal adaptation, is the product of blind variation and selective retention—trial and error. The same process governs both biological emergence and the growth of knowledge in science. (Bartley, 1987b, p. 20)

Evolutionary epistemology does have a certain prima facie attractiveness, but it also has a number of problems. Ernest Gellner (1985, pp. 46-52) provides a discussion of many of these problems and criticizes evolutionary epistemology—he calls it the "continuity thesis"—from within the Popperian tradition. At this point, all that can be said is that evolutionary epistemology is a possible solution to Popper's problem; time will tell if it can be worked into a fully adequate solution.

It should also be noted that evolutionary epistemology is a much broader program than the Popperian version of it mentioned here. Hull (1988) and Rescher (1990) present alternative, not necessarily Popperian, versions of evolutionary epistemology. The field is sufficiently broad that Campbell, Heyes, and Callebaut (1987) have been able to provide a 600-item bibliography of it.

32. There are a number of questions that need to be sorted out regarding the relationship between Rorty's own pragmatic-hermeneutic view and CR-cum-evolutionary epistemology. For one thing, Rorty claims Dewey as inspi-

ration, but so does Bartley (1987a, p. 213; 1990, p. 241). For another thing, the "naturalized epistemology" of Quine and others plays an important role in Rorty's argument, and yet evolutionary epistemology looks a great deal like biologically naturalized epistemology. It should be noted that a similar Popper-based critique of Rorty is given in Backhouse (1991a and 1991b).

33.   One is reminded of Morris Zapp's comment to Fulvia Morgana in David Lodge's *Small World:*

> Well, I'm a bit of a deconstructionist myself. It's kind of exciting—the last intellectual thrill left. Like sawing through the branch you're sitting on. (Lodge, 1984, p. 118)

34.   See Habermas (1981, 1987a, and 1987b). Rorty, for example—after discussing what he calls know-nothing critiques of antimodernism—says,

> But the same accusations are made by writers who know what they are talking about, and whose views are entitled to respect. As I have already suggested, the most important of these writers is Habermas, who has mounted a sustained, detailed, carefully argued polemic against critics of the Enlightenment (e.g. Adorno, Foucault), who seem to turn their back on the social hopes of liberal societies. (Rorty, 1989b, p. 82)

35.   This quote is then followed by criticism: "I think that critical rationalism, by clinging to the idea of irrefutable roles of criticism, allows a weak version of the Kantian justificatory mode to sneak into its inner precincts through the back door" (Habermas, 1987a, p. 302). The point being made in the text is *not* that Habermas supports CR: *he does not*. In fact Habermas's work has inspired an alternative social approach to the growth of knowledge: the "finalizationists" (see Fuller, 1988, pp. 183-84). The point is simply that Habermas considers CR to be an attempt to solve the same problem—the critique of radical postmodernism—that his own theory of communicative action is designed to solve.

36.   The term *thick methodology* was used by Donald McCloskey (1988a) to characterize his own rhetoric of economics. My argument—more in the spirit of criticisms by A. W. Coats (1988) and Uskali Mäki (1991b)—is that, while one *could* conduct the rhetoric of economics in this way, this is *not* the approach that McCloskey himself takes.

37.   For McCloskey there is no inconsistency here; both views basically argue that "the higher authority"—philosophy in the case of Rorty, and the government in the case of Chicago economics—has no business meddling into our affairs. The problem is that in Rorty's view, or any other Wittgensteinian-based view of human behavior, we are forever trapped in a framework (language game, culture, discourse community, etc.): our point of view holds us; we do not hold it (Fish, 1988, p. 27). Such a view is clearly at odds with the Chicago view of choice, volition, and agency, and any ethical theory based on such notions. This latter point—the inconsistency between McCloskey's

Chicago view of individual agency and his socially constructivist view of the communicative order—has been argued quite persuasively by Mirowski (1987 and 1992). The relationship between McCloskey's views on the rhetoric of economics and his commitment to classical liberalism is also discussed in Evensky (1992).

38. Or more geographically: the Chicago, Virginia, and University of Washington schools (this latter label is due to North, 1990, p. 27). What Bartley would exclude from the discussion of knowledge are fields such as the economics of booms and slumps (i.e., macroeconomics), econometrics, general equilibrium theory, and of course Marx (Bartley, 1990, p. 91).

It is useful to note that Bartley's interpretation of these matters provides a way of reconciling Popper's suggestion that his approach to the growth of knowledge is simply an application of situational analysis and thus neoclassical economic theory (see Chapter 6 of this book).

39. This type of argument motivates work such as Diamond (1988b) and Radnitzky (1986 and 1989). I would like to note that I became aware of Wible (1992) too late to include it in my discussion but I suspect that it will also be an important contribution to this topic.

40. The Poverty of Historicism (Popper, 1961) originally appeared in three issues of Economica in 1944 and 1945.

41. Ernest Gellner makes a similar point: "*Which* sociology is the philosophy of science to use? Which sociological paradigm may we trust, when using sociology to grapple with the general problem of the nature of science, so as to illuminate the standing of all sciences, including sociology itself?" (1985, p. 110).

42. I personally find this quagmire rather fascinating; it is just beyond the scope of the current project.

43. "May" is used here because some have argued (even some Popperians) that Bartley's comprehensively critical rationalism contains its own type of circularity (e.g., Hauptli, 1991; Watkins, 1969, 1971, and 1987).

44. This is a rather frequent criticism of the sociology of knowledge. For example, Bernard Susser argues,

Sociological relativism is intrinsically untenable because, predictably, it entails its own self-destruction. As a consistent system of ideas, the sociology of knowledge cannot refrain from asking about its own genealogy and its own mundane entanglements. Once these are acknowledged, the sociology of knowledge is either dismissed as an evanescent, culture-based intellectual fashion or it contradicts (and disqualifies) its assumptions by claiming for itself an Archimedian cognitive position. (Susser, 1989, p. 248)

45. Mary Hesse (1988, p. 101), for example, argues that this purported circularity is not vicious. On the other hand, Robert Nola (1990) suggests there are a number of cases of circularity and regress within the strong program. It should also be noted that sociologists of scientific knowledge are not unaware of this problem; it is called the "problem of reflexivity," and a number

of authors—Steve Woolgar, in particular—have attempted to circumvent it. It remains an open question how successful such efforts have been.

46. Caldwell (1982) advocated a pluralist position, but this view was not linked explicitly to the Bartley interpretation of Popper until later work—especially Caldwell (1991a). This is also one way to read Lawrence Boland's extensive writings on the methodology of economics (Boland, 1982, 1986, and 1989, for example).

David Colander (1989 and 1990) presents a view that has much in common with the views of Caldwell and Redman but without any reference to Bartley or the related Popperian tradition; unfortunately Colander calls his view the "sociological approach to methodology" (1990, p. 191).

47. Patricia S. Churchland (1986, pp. 290-91) cites an interesting laboratory study where college students were asked to predict and explain the behavior of certain moving physical objects. Surprisingly, most of the students were intuitively more Aristotelian than Newtonian, and some even provided rather elaborate explanations for the failure of their predictions.

These experiments indicate that apart from the scientific community, most humans seem to employ a sort of *folk physics*. . . . Although a person might not describe his set of beliefs concerning motion as a theory, given the role of such beliefs in explaining and predicting motion, it is nonetheless appropriate to credit the person with a folk theory. (Churchland, P. S., 1986, p. 290)

48. "Intentional causality" is John Searle's term (e.g., 1991, p. 335); this is not the way that Rosenberg and many others would characterize the issue. As we will see below, precisely how one causally connects the giving of reasons with the action in the explanandum (as in [L]) is in fact a major point of contention in folk psychology literature.

49. This argument basically combines thesis 10 and thesis 12 from the thirteen theses of Chapter 10.

50. Obviously, not all of (even neoclassical) economics is microeconomics and, as I argued in Hands (1991b), there is more to even microeconomics than the theory of individual behavior (also see Nelson 1990). The question of how much neoclassical microeconomics is about individual behavior, and how much it is about aggregative market phenomena is a topic that needs additional consideration. Important recent studies (such as Mirowski, 1989) that emphasize the influence of energy physics on the development of neoclassical economics seem to focus almost exclusively on aggregated market phenomena and neglect agency and the microeconomic theory of individual choice. Thus it appears there are (at least) two separate strains of neoclassical economics: one that focuses on the market where the individual atoms are indistinguishable and devoid of anything that might be called agency, and a second strain that is concerned with explaining the actions of individual economic agents. The former—basically, Walrasian general equilibrium theory—seems to have drawn on physics (particularly energy physics) for its scientific inspiration and legitimacy (and has generated its own particular set of prob-

lems), while the latter—neoclassical microeconomics as a theory of individual behavior—struggles with the issues discussed above. As I say, these issues need a more careful examination. At the very least it seems that "neoclassical economics" is not a "natural kind."

51. One of the main themes of behaviorist psychology was the rejection of folk psychology. Stephen Stich explains the behaviorist position in the following way.

> Common sense offers *explanations:* people act as they do because they have certain beliefs or wants or fears or desires. But if behavioristic psychology was on the right track, then these explanations were mistaken. People act as they do because they have been subjected to certain histories of reinforcement, and the beliefs and desires, hopes and fears of common sense psychology have nothing to do with their behavior. These mental states of folk psychology are, on the behaviorists' view, no more than myths conjured by primitive theory. Like the deities, humors, and vital forces of other primitive theories, they simply do not exist. Our readiness to describe ourselves in these mythical terms indicates only that we are still wedded to a wrong-headed prescientific theory, just as in earlier centuries people under the spell of other primitive theories described themselves and their fellows as bewitched or phlegmatic. (Stich, 1983, p. 2)

52. Churchland, P. S. (1986), for example. See note 58 below.

53. This is *not* to say that all of the literature *about* folk psychology and intentional explanations is *against* folk psychology and intentional explanations. Quite the contrary: much of what has recently been written about folk psychology has defended it. But in a sense this only proves the point. If advocates of folk psychology now feel obligated to write spirited books in its defense, the threat must be serious. Remember, folk psychology essentially says that we do things because we have reasons for doing them; if a pretty big storm were not brewing over our basic intuitions about human behavior, it seems highly unlikely that such a commonsense notion would require a philosophical "defense."

54. Note the emphasis on a *scientific* theory of human behavior. Even those authors who are most hostile to folk psychology are not campaigning to remove belief and desire from our everyday vocabulary or even our everyday explanations of ourselves and others. The criticism is directed at folk psychology as the basis for a *scientific* theory of human behavior. In fact, one response to the problems of folk psychology is social science instrumentalism—a position usually associated with Daniel Dennett (1978, 1987). The argument is that, while social sciences that employ belief-and-desire explanations do not provide causal explanations, they may continue to be instrumentally useful in practical applications. A similar argument is made in Stephen Stich's "modified Panglossian view" where the generalizations of folk psychology are given the same status as the "culinary generalizations" used in cooking. "When suitably hedged, the economists' or sociologists' generalizations may be both true and useful, just as the chef's are" (Stich, 1983, p. 228).

55. As Rosenberg argues:

> Now there is doubtless a good explanation of why we have become attached to the kind-terms *action, desire,* and *belief* as the explanatory variables for human behavior. They have emerged as tools for guiding our expectations about how others will act, but we have uncovered no laws about the behavior they explain. Perhaps the failure to find laws about this behavior is the result of the fact that these kind-terms are not "natural," they just don't carve things up at the joints. Like our example "fish," every generalization that employs them is so riddled with exceptions that there are no laws we can discover to be stated in these terms. (Rosenberg, 1988a, pp. 12-13)

56. A small sample of this literature starting with the most pro-intentionality and moving steadily toward the other extreme of eliminative materialism would be the following: Searle (1983, 1991), Lennon (1990), Dretske (1988), Dennett (1987), Stich (1983), and Churchland, P. S. (1986). Although arguments could be made for switching some of the authors in the middle of this group, the endpoints seem to be firmly in place.

57. It is not clear that Rosenberg has continued to advocate such a strong eliminativist position in later work.

58. While my separation of this discussion of eliminative naturalism from the previous discussion of antimodernism may suggest that the two topics are entirely unrelated, this is not really the case. In many respects the two literatures are rather closely related. Most obviously, if eliminative materialism is unconditionally accepted and the entire notion of "belief" is relegated to the same status as demons and goblins, *then traditional epistemology is automatically eliminated.* Traditional epistemology is concerned with which beliefs are justified, which beliefs can be given foundations; if the entire concept of belief is eliminated, then surely the question of which such nonexistent thing one ought to hold goes with it.

This elimination is not too surprising once one realizes that much of the philosophical inspiration for eliminative materialism comes from Quine's position that there is no "first philosophy": there is no place to stand outside of science in order to evaluate science. According to Quine's naturalized epistemology, we have only science; and therefore if we are to investigate the class of issues that have traditionally been associated with epistemology, we must begin with our best science (including possibly neuroscience). Such a view points quite naturally down the road to eliminative materialism. Now, Rorty also argues that there is no first philosophy; but his view—and that of antimodernism more generally—is often taken to undermine science. The difference is that, *if* science receives its primary support from the epistemological stories about its access to privileged representations, then undermining that epistemology as Rorty does also undermines the privileged position of science and brings it down to the level of any other (all other) discourse. On the other hand, *if* science is self-supporting (either prima facie or because of its instrumental success) as I would characterize the Quinean view, then the destruction of the narrative about its epistemic privilege is only a problem for philosophy; it leaves science unscathed.

Clearly, the authors that are closest to what I have labeled "radical post-modernism" (mostly Continental) intend to bring science down. For other, less radical, antimodernists there does not seem to be any general consensus on this particular issue. This ambiguity is particularly true of Feyerabend and Rorty. Both of these influential antimodernist authors have written papers that defend eliminative materialism (Feyerabend, 1963; and Rorty 1965, 1970, and 1982b); in fact, Rorty (1970) actually has the title "In Defense of Eliminative Materialism." Despite these early contributions to the literature, it is not clear that the two authors would continue to agree (or ever did agree, for that matter) on the question of eliminative materialism. Let me briefly consider what seems to be their current views.

In the case of Feyerabend, it is clear that he now disavows his earlier position, or at least the part of it that might conflict with his antimodernism. The following note was added to the 1981 reprint of his 1963 paper.

> I no longer agree with the assumption . . . that the "correctness" of an idiom, or of the statements that can be formulated in its terms, empirical statements included, is independent of the (linguistic) practice to which the statements belong: the truth, even of "empirical" statements, may be *constituted* by the fact that they are part of a certain form of life which assembles evidence in a certain way. "Athene has provided me with new strength" can be a true observation statement for a Homeric warrior. . . . Secondly, the choice of an idiom for the description of mental events cannot be decided by considerations of testability and "cognitive content" alone. It may well be that a materialistic language (if it ever gets off the ground) is richer in cognitive content than common sense. . . . But it will be much poorer in other respects. For example, it will lack the associations which now connect mental events with emotions, our relations to others, and which are the basis of the arts and the humanities. We therefore have to make a choice: do we want scientific efficiency, or do we want a rich human life of the kind now known to us and described by our artists? The choice concerns *the quality of our lives*—it is a *moral* choice. (Feyerabend, 1981, p. 163)

In the case of Rorty, his current position seems less clear. While most commentators on Rorty's anti-epistemology have simply avoided his writings on materialism and the philosophy of mind, Jennifer Hornsby (1990) argues that *Philosophy and the Mirror of Nature* is not only consistent with these writings but actually represents a radical form of eliminative materialism. Her argument is that according to Rorty the dualist metaphor of the mirror directs our attention to the "problem of the mind"—the problem of the "nature of the mind" as a philosophical problem. Earlier materialists argued that the mind could be reduced to physical processes, but they were still "trapped" in dualism since they allowed dualist notions to continue to frame the question. With the more radical interpretation of Rorty, dualism is simply abandoned—

lock, stock, and barrel—and so the problem of the mind disappears with it. For Rorty, in Hornsby's interpretation, "mental phenomena have been precluded from the picture before a question about their location can arise" (Hornsby, 1990, p. 53). I believe that this interpretation of Rorty holds up as long as one characterizes "materialism" as Rorty would, as that which is described by the language of science—"materialistic science (and what other kind of science is there?)" (Rorty, 1982b, p. 345)—rather than as the ontological materialism that most people tend to associate with the word *materialism*. For Rorty the concern is solely with the signifiers and not with the signified; "there is no such thing as comparing a linguistic formulation with a bit of non-linguistic knowledge, but only a matter of seeing how various linguistic items fit together with other linguistic items and with the purposes for which language as a whole is to be used" (1982b, p. 344). In defending the "materialist" language, Rorty says,

> I am merely claiming the same legitimacy for the neurological vocabulary— where "legitimacy" means the right to be considered a report of experience. My attitude is not that some vocabularies are "legitimate," but rather we should let a thousand vocabularies bloom and then see which survive. (Rorty, 1970, p. 119)

59. Popper (1977) and various parts of Popper (1972 and 1982) in particular.

60. One needs to be careful about the term *reduction*. As Popper seems to be using it here and as it is often used in informal conversation, reduction means something like "reduction and therefore elimination." The problem is that eliminative reduction is only one very special type of reduction, and it is not the way that intertheoretic reduction has traditionally been defined, as in Nagel (1961, ch. 11). In fact, in much of the eliminative materialist literature, the argument is made that belief and desire *cannot be reduced* to brain processes and this is exactly why they need to be *eliminated*. For a recent discussion of reduction in general, see Kuipers (1990); and for a discussion of how it specifically relates to the question of eliminative materialism, see Schwartz (1991).

61. I discuss Popper's relationship to the D-N model in more detail in Hands (1991b).

62. In Hands (1991b), I argued that there were effectively three Popperian solutions to this tension: the verisimilitude solution, the Lakatosian solution, and the critical rationalist solution. Arguments in Chapters 8 and 9 of this book should show that verisimilitude is not a viable option, while arguments in Chapters 5, 6, and 8 should demonstrate that Lakatos also fails to provide a solution; only critical rationalism remains. I also argued in Hands (1991b) that one could release the tension by weakening the notion of an acceptable explanation from the covering law model to a more pragmatic notion of explanation, but this solution cannot really be classified as Popperian, given Popper's adherence to the D-N model and deductivism more generally.

In Caldwell (1991c), the explicitly Popperian CR of Caldwell (1991a) is expanded into a slightly more general position called "critical pluralism." This critical pluralism draws on the work of Popper and Bartley as well as non-Popperian philosophers such as Larry Laudan. Since the issue at hand is to characterize the best *Popperian* response to antimodernism and eliminative naturalism, I will focus exclusively on the self-consciously Popperian arguments in Caldwell (1991a) rather than on critical pluralism more generally. This is not intended as a criticism of critical pluralism, but rather as an attempt to avoid introducing yet another issue: the question of the degree to which critical pluralism is "Popperian."

63. See the fourth criticism of folk psychology discussed above in the text, as well as notes 55 and 60.

64. It is at odds with materialism, however, if one treats Popper's "three worlds" story as a serious story about ontology. If it is really offered as an ontology, then I would suggest that it is adventitious to main themes of the Popperian tradition, and that anything the Popperian tradition has to say about anti-inductivism, fallibilism, testing, rationality, or criticism can fit quite comfortably with a materialist ontology.

65. With the caveat in note 64.

66. See note 31 above.

67. Bruce Caldwell seems to lean this way in Caldwell (1991a and 1991b).

68. There seems to be yet another version of the Popperian program: Anderson's "refined falsificationism" (see Radnitzky, 1991).

References

Adler, J. E. and Elgin, C. Z. 1980. "Review of Imre Lakatos' *Philosophical Papers.*" *Synthese,* 43, 411-20.

Agassi, J. 1960. "Methodological Individualism." *British Journal of Sociology,* 11, 244-70.

Agassi, J. 1975. "Institutional Individualism." *British Journal of Sociology,* 26, 144-55.

Agassi, J. 1979. "The Legacy of Lakatos." *Philosophy of the Social Sciences,* 9, 316-26.

Agassi, J. 1988. *The Gentle Art of Philosophical Polemics.* LaSalle, IL: Open Court.

Agassi, J. and Klappholz, K. 1959. "Methodological Prescriptions in Economics." *Economica,* 26, 60-74.

Agassi, J. and Klappholz, K. 1960. "A Rejoinder." *Economica,* 26, 160-61.

Ahonen, G. 1989. "On the Empirical Content of Keynes' *General Theory.*" *Ricerche Economiche,* 43, 256-69.

Ahonen, G. 1990. "Commentary on Hands' 'Second Thoughts on "Second Thoughts"' on the Lakatosian Progress of *The General Theory.*" *Review of Political Economy,* 2, 94-101.

Albert, H. 1985. *Treatise on Critical Reason.* Princeton, NJ: Princeton University Press.

Amariglio, J. 1990. "Economics as Postmodernist Discourse." In W. J. Samuels, ed., *Economics as Discourse.* Boston: Kluwer Academic, 15-46.

Anderson, G. 1986. "Lakatos and Progress and Rationality in Science: A Reply to Agassi." *Philosophia,* 16, 239-43.

Archibald, G. C. 1979. "Method and Appraisal in Economics." *Philosophy of the Social Sciences,* 9, 304-15.

Ariew, R. 1984. "The Duhem Thesis." *British Journal for the Philosophy of Science,* 35, 313-25.

Arrow, K. J. and Hahn, F. H. 1971. *General Competitive Analysis.* San Francisco: Holden Day.

Backhouse, R. E. 1991a. "Pragmatism, Post-modernism, and Literary Theory: On Some Recent Challenges to Economic Methodology." Paper presented at the History of Economics Society annual meetings, College Park, MD, June 1991.

Backhouse, R. E. 1991b. "Rhetoric and Methodology." University of Birmingham Discussion Papers in Economics, 91-10.

Backhouse, R. E. 1991c. "The Constructivist Critique as Economic Methodology." University of Birmingham Discussion Papers in Economics, 91-12.

Backhouse, R. E. 1992. "How Should We Approach the History of Economic Thought, Fact, Fiction, or Moral Tale?" *Journal of the History of Economic Thought,* 14, 18-35.

Blazer, W. 1982. "A Logical Reconstruction of Pure Exchange Economics." *Erkenntnis,* 15, 33-53.

Barnes, B. 1974. *Scientific Knowledge and Sociological Theory.* London: Routledge and Kegan Paul.

Barnes, B. 1977. *Interests and the Growth of Knowledge.* London: Routledge and Kegan Paul.

Barro, R. and Grossman, H. 1976. *Money, Employment, and Inflation.* Cambridge, MA: Harvard University Press.

Bartley, W. W. III. 1982. "The Philosophy of Karl Popper: Part III, Rationality, Criticism, and Logic." *Philosophia,* 11, 121-221.

Bartley, W. W. III. 1984. *The Retreat to Commitment,* 2nd edition. LaSalle, IL: Open Court; 1st edition, 1962.

Bartley, W. W. III. 1987a. "Theories of Rationality." In G. Radnitzky and W. W. Bartley III, eds., *Evolutionary Epistemology, Rationality, and the Sociology of Knowledge.* LaSalle, IL: Open Court, 205-14.

Bartley, W. W. III. 1987b. "Philosophy of Biology versus Philosophy of Physics." In G. Radnitzky and W. W. Bartley III, eds., *Evolutionary*

Epistemology, Rationality, and the Sociology of Knowledge. LaSalle, IL: Open Court, 7-45.

Bartley, W. W. III. 1990. *Unfathomed Knowledge, Unmeasured Wealth.* LaSalle, IL: Open Court.

Baynes, K., Bohman, J., and McCarthy, T., eds. 1987. *After Philosophy: End or Transformation?* Cambridge, MA: MIT Press.

Begg, D. K. H. 1980. "Rational Expectations and the NonNeutrality of Systematic Monetary Policy." *Review of Economic Studies,* 47, 293-303.

Begg, D. K. H. 1982. *The Rational Expectations Revolution in Macroeconomics.* Baltimore: Johns Hopkins University Press.

Blaug, M. 1975. "Kuhn versus Lakatos, or Paradigms versus Research Programmes in the History of Economics." *History of Political Economy,* 7, 399-419.

Blaug, M. 1976a. "Kuhn versus Lakatos, or Paradigms versus Research Programmes in the History of Economics." In S. J. Latsis, ed., *Method and Appraisal in Economics.* Cambridge, England: Cambridge University Press, 149-80.

Blaug, M. 1976b. "The Empirical Status of Human Capital Theory: A Slightly Jaundiced Survey." *Journal of Economic Literature,* 14, 827-55.

Blaug, M. 1978. *Economic Theory in Retrospect,* 3rd edition. Cambridge, England: Cambridge University Press.

Blaug, M. 1980a. *The Methodology of Economics.* Cambridge, England: Cambridge University Press.

Blaug, M. 1980b. *A Methodological Appraisal of Marxian Economics.* Amsterdam: North-Holland.

Blaug, M. 1985. "Comments on D. Wade Hands, 'Karl Popper and Economic Methodology: A New Look.'" *Economics and Philosophy,* 1, 286-88.

Blaug, M. 1987. "Second Thoughts on the Keynesian Revolution." Unpublished English version of "Ripensamenti Sulla Rivoluzione Keynesiana." *Rassegna Economica,* 51, 605-34. (See Blaug, 1991.)

Blaug, M. 1990. "Reply to D. Wade Hands' 'Second Thoughts on "Second Thoughts"': Reconsidering the Lakatosian Progress of *The General Theory.*'" *Review of Political Economy,* 2, 102-4.

Blaug, M. 1991. "Second Thoughts on the Keynesian Revolution." *His-*

tory of Political Economy, 23, 171-92. (Published version of Blaug, 1987.)

Blaug, M. 1992. "Comment on D. W. Hands' 'Falsification, Situational Analysis, and Scientific Programmes: The Popperian Tradition in Economic Methodology.'" In N. de Marchi, ed., *Post-Popperian Methodology of Economics.* Boston: Kluwer Publishing.

Blaug, M. and de Marchi, N., eds. 1991. *Appraising Modern Economics: Studies in the Methodology of Scientific Research Programmes.* Aldershot, England: Edward Elgar.

Block, W. 1980. "On Robert Nozick's 'On Austrian Methodology.'" *Inquiry,* 23, 397-444.

Bloor, D. 1976. *Science and Social Imagery.* London: Routledge and Kegan Paul.

Bloor, D. 1984. "The Strengths of the Strong Programme." In J. R. Brown, ed., *Scientific Rationality: The Sociological Turn.* Dordrecht, Holland: D. Reidel, 75-94.

Bloor, D. 1992. "Left and Right Wittgensteinians." In A. Pickering, ed., *Science as Practice and Culture.* Chicago: University of Chicago Press, 266-82.

Boland, L. A. 1979. "A Critique of Friedman's Critics." *Journal of Economic Literature,* 17, 503-22.

Boland, L. A. 1981. "On the Futility of Criticizing the Neoclassical Maximization Hypothesis." *American Economic Review,* 71, 1031-36.

Boland, L. A. 1982. *The Foundations of Economic Method.* London: Allen and Unwin.

Boland, L. A. 1983. "The Neoclassical Maximization Hypothesis: Reply." *American Economic Review,* 73, 828-30.

Boland, L. A. 1986. *Methodology for a New Microeconomics.* Boston: Allen and Unwin.

Boland, L. A. 1989. *The Methodology of Economic Model Building.* London: Routledge.

Boland, L. A. and Frazer, W. J. 1983. "An Essay on the Foundations of Friedman's Methodology." *American Economic Review,* 73, 129-44.

Brown, E. K. 1981. "The Neoclassical and Post-Keynesian Research Programs: The Methodological Issues." *Review of Social Economy,* 34, 111-32.

Brown, J. R. 1980. "History and the Norms of Science." In P. D. Asquith and R. N. Giere, eds., *PSA 1981*, Volume 1. East Lansing, MI: PSA, 236-48.

Brush, S. C. 1974. "Should the History of Science Be Rated X?" *Science*, 183, 1164-72.

Burkhardt, J. and Canterbery, E. 1986. "The Orthodoxy and Professional Legitimacy: Toward a Critical Sociology of Economics." *Research in the History of Economic Thought and Methodology*, 4, 229-50.

Caldwell, B. J. 1980. "A Critique of Friedman's Methodological Instrumentalism." *Southern Economic Journal*, 47, 366-74.

Caldwell, B. J. 1981. "Review of *The Methodology of Economics* by Mark Blaug." *Southern Economic Journal*, 48, 242-45.

Caldwell, B. J. 1982. *Beyond Positivism: Economic Methodology in the Twentieth Century*. London: Allen and Unwin.

Caldwell, B. J. 1983a. "Hayek the Falsificationist? A Rebuttal to Hutchison on Hayek's Methodology." Paper presented at the History of Economics Society Annual Meetings, Charlottesville, VA, May 1983.

Caldwell, B. J. 1983b. "The Neoclassical Maximization Hypothesis: Comment." *American Economic Review*, 73, 824-27.

Caldwell, B. J. 1984a. "Some Problems with Falsificationism in Economics." *Philosophy of the Social Sciences*, 14, 489-95.

Caldwell, B. J. 1984b. "Praxeology and Its Critics: An Appraisal." *History of Political Economy*, 16, 363-79.

Caldwell, B. J. 1988a. "The Case for Pluralism." In N. de Marchi, ed., *The Popperian Legacy in Economics and Beyond*. Cambridge, England: Cambridge University Press, 231-44.

Caldwell, B. J. 1988b. "Clarifying Popper." Paper presented at the History of Economics Society Annual Meetings, Richmond, VA, June 1988. (Early draft of Caldwell, 1991a.)

Caldwell, B. J. 1990. "Does Methodology Matter? How Should It Be Practiced?" *Finnish Economic Papers*, 3, 64-71.

Caldwell, B. J. 1991a. "Clarifying Popper." *Journal of Economic Literature*, 29, 1-33.

Caldwell, B. J. 1991b. "Friedman's Predictivist Instrumentalism: A Modification." In *Research in the History of Economic Thought and Methodology* (forthcoming).

Caldwell, B. J. 1991c. "The Methodology of Scientific Research Programmes in Economics: Criticisms and Conjectives." In G. K. Shaw, ed., *Economics, Culture, and Education: Essays in Honour of Mark Blaug.* Aldershot, England: Edward Elgar, 95-107.

Caldwell, B. J. and Coats, A. W. 1984. "The Rhetoric of Economists: A Comment on McCloskey." *Journal of Economic Literature,* 22, 575-78.

Campbell, D. T., Heyes, C. M., and Callebaut, W. G. 1987. "Evolutionary Epistemology Bibliography." In W. Callebaut and R. Pinxten, eds., *Evolutionary Epistemology: A Multiparadigmatic Program.* Dordrecht, Holland: D. Reidel, 405-31.

Caputo, J. D. 1983. "The Thought of Being and the Conversation of Mankind: The Case of Heidegger and Rorty." *Review of Metaphysics,* 36, 661-85.

Carrier, M. 1988. "On Novel Facts: A Discussion of Criteria for Non-ad-hoc-ness in the Methodology of Scientific Research Programmes." *Zeitschrift fur Allgemeine Wissenschaftstheorie,* 19, 205-231.

Chipman, J. S. 1966. "A Survey of the Theory of International Trade: Part 3, The Modern Theory." *Econometrica,* 34, 18-76.

Churchland, P. M. 1984. *Matter and Consciousness.* Cambridge, MA: MIT Press.

Churchland, P. S. 1986. *Neurophilosophy: Toward a Unified Science of the Mind-Brain.* Cambridge, MA: MIT Press.

Clower, R. 1965. "The Keynesian Counter-revolution." In F. H. Hahn and F. Brechling, eds., *The Theory of Interest Rates.* London: Macmillan, 103-25.

Clower, R. 1975. "Reflections on the Keynesian Perplex." *Zeitschrift fur Nationalokonomie,* 35, 1-24.

Coats, A. W. 1976. "Economics and Psychology: The Death and Resurrection of a Research Programme." In S. J. Latsis, ed., *Method and Appraisal in Economics.* Cambridge, England: Cambridge University Press, 43-64.

Coats, A. W. 1984. "The Sociology of Knowledge and the History of Economics." *Research in the History of Economic Thought and Methodology,* 2, 211-34.

Coats, A. W. 1988. "Economic Rhetoric: The Social and Historical Context." In A. Klamer, D. N. McCloskey, and R. M. Solow, eds., *The Consequences of Economic Rhetoric.* Cambridge, England: Cambridge University Press, 64-84.

Coats, A. W. and Pressman, S. 1987. "The Rhetoric of Economics: Further Comments." *Eastern Economic Journal,* 13.

Coddington, A. 1972. "Positive Economics." *Canadian Journal of Economics,* 5, 1-15.

Coddington, A. 1975. "The Rationale of General Equilibrium Theory." *Economic Inquiry,* 13, 539-58.

Cohen, A. 1983. "The Laws of Return Under Competitive Conditions: Progress in Microeconomics since Sraffa." *Eastern Economic Journal,* 9.

Cohen, A. and Dascal, M. 1989. *The Institution of Philosophy: A Discipline in Crisis?* LaSalle, IL: Open Court.

Colander, D. 1989. "The Invisible Hand of Truth." In D. Colander and A. W. Coats, eds., *The Spread of Economic Ideas.* Cambridge, England: Cambridge University Press, 31-36.

Colander, D. C. 1990. "Workmanship, Incentives, and Cynicism." In A. Klamer and D. Colander, eds., *The Making of an Economist.* Boulder, CO: Westview Press, 187-200.

Collins, H. 1985. *Changing Order: Replication and Induction in Scientific Practice.* Los Angeles, CA: Sage.

Collins, H. 1991. "The Meaning of Replication and the Science of Economics." *History of Political Economy,* 23, 123-42.

Cross, R. 1982. "The Duhem-Quine Thesis, Lakatos, and the Appraisal of Theories in Macroeconomics." *Economic Journal,* 92, 320-40.

Dascal, M. 1989. "Reflections on the 'Crisis of Modernity.'" In A. Cohen and M. Dascal, eds., *The Institution of Philosophy.* LaSalle, IL: Open Court, 217-40.

Davis, J. B. 1990. "Rorty's Contribution to McCloskey's Understanding of Conversation as the Methodology of Economics." *Research in the History of Economic Thought and Methodology,* 7, 73-85.

de Marchi, N. 1976. "Anomaly and the Development of Economics: The Case of the Leontief Paradox." In S. J. Latsis, ed., *Method and Appraisal in Economics.* Cambridge, England: Cambridge University Press, 109-27.

de Marchi, N., ed. 1988a. *The Popperian Legacy in Economics and Beyond.* Cambridge, England: Cambridge University Press.

de Marchi, N. 1988b. "Popper and the LSE Economists." In N. de Marchi, ed., *The Popperian Legacy in Economics and Beyond.* Cambridge, England: Cambridge University Press, 139-66.

de Marchi, N., ed. 1992. *Post-Popperian Methodology of Economics.* Boston: Kluwer Publishing.

de Marchi, N. and Gilbert, C., eds. 1989. "History and Methodology in Econometrics." *Oxford Economic Papers,* 41.

Debreu, G. 1983. *Mathematical Economics: Twenty Papers of General Debreu.* Cambridge, England: Cambridge University Press.

Dennett, D. 1978. *Brainstorms: Philosophical Essays on Mind and Psychology.* Cambridge, MA: MIT Press.

Dennett, D. 1982. "Comments on Rorty." *Synthese,* 53, 349-56.

Dennett, D. 1987. *The Intentional Stance.* Cambridge: MA: MIT Press.

Diamond, A. M., Jr. 1988a. "The Empirical Progressiveness of the General Equilibrium Research Program." *History of Political Economy,* 20, 119-35.

Diamond, A. M., Jr. 1988b. "Science as a Rational Enterprise." *Theory and Decision,* 24, 147-67.

Dretske, F. 1988. *Explaining Behavior: Reasons in a World of Causes.* Cambridge, MA: MIT Press.

Duhem, P. 1954. *The Aim and Structure of Physical Theory,* translated by P. P. Wiener. Princeton, NJ: Princeton University Press.

Dyer, A. W. 1988. "Economic Theory as Art Form." *Journal of Economic Issues,* 22, 157-66.

Evensky, J. 1992. "Ethics and the Classical Liberal Tradition in Economics." *History of Political Economy,* 24, 61-77.

Farr, J. 1983. "Popper's Hermeneutics." *Philosophy of the Social Sciences,* 13, 157-76.

Faulhaber, G. R. and Baumol, W. J. 1988. "Economists as Innovators: Practical Products of Theoretical Research." *Journal of Economic Literature* 26, 577-600.

Feyerabend, P. K. 1963. "Materialism and the Mind-Body Problem." *Review of Metaphysics,* 17 [reprinted in Feyerabend (1981)].

Feyerabend, P. K. 1970. "Consolations for the Specialist." In I. Lakatos and A. Musgrave, eds., *Criticism and the Growth of Knowledge.* Cambridge, England: Cambridge University Press, 197-230.

Feyerabend, P. K. 1974. "Popper's Objective Knowledge." *Inquiry,* 17, 475-507.

Feyerabend, P. K. 1975a. *Against Method.* London: New Left Books.

Feyerabend, P. K. 1975b. "Imre Lakatos." *British Journal for the Philosophy of Science,* 26, 1-18.

Feyerabend, P. K. 1976. "On the Critique of Scientific Reason." In R. S. Cohen et al., eds., *Essays in Memory of Imre Lakatos.* Dordrecht, Holland: D. Reidel, 109-43.

Feyerabend, P. K. 1981. *Realism, Rationalism, and Scientific Method: Philosophical Papers,* Volume 1. Cambridge, England: Cambridge University Press.

Fish, S. 1988. "Comments from Outside Economics." In A. Klamer, D. N. McCloskey, and R. M. Solow, eds., *The Consequences of Economic Rhetoric.* Cambridge, England: Cambridge University Press, 21-30.

Fisher, F. M. 1976. "The Stability of General Equilibrium: Results and Problems." In M. J. Artis and A. R. Nobay, eds., *Essays in Economics and Analysis.* Cambridge, England: Cambridge University Press, 3-29.

Fisher, F. M. 1983. *Disequilibrium Foundations of Equilibrium Economics.* Cambridge, England: Cambridge University Press.

Fisher, R. 1986. *The Logic of Economic Discovery.* New York: New York University Press.

Friedman, M. 1953. "The Methodology of Positive Economics." In *Essays in Positive Economics.* Chicago: University of Chicago Press, 3-43.

Fuller, S. 1988. *Social Epistemology.* Bloomington, IN: Indiana University Press.

Fulton, G. 1984. "Research Programmes in Economics." *History of Political Economy,* 16, 187-205.

Gale, D. 1965. "A Note on Global Instability of Competitive Equilibrium." *Naval Research Logistics Quarterly,* 10, 80-87.

Galison, P. 1987. *How Experiments End.* Chicago: University of Chicago Press.

Gardner, M. R. 1982. "Predicting Novel Facts." *British Journal for the Philosophy of Science,* 33, 1-15.

Garfinkel, H. 1967. *Studies in Ethnomethodology.* Englewood Cliffs, NJ: Prentice Hall.

Garfinkel, H., Lynch, M., and Livingston, E. 1981. "The Work of a Discovering Science Construed with Materials from the Optimally Discovered Pulsar." *Philosophy of the Social Sciences,* 11, 131-58.

Gellner, E. 1974. *Legitimation of Belief.* Cambridge, England: Cambridge University Press.

Gellner, E. 1985. *Relativism and the Social Sciences.* Cambridge, England: Cambridge University Press.

Gibbard, A. and Varian, H. R. 1978. "Economic Models." *Journal of Philosophy,* 75, 664-77.

Goodwin, C. 1980. "Towards a Theory of the History of Economics." *History of Political Economy,* 12, 610-19.

Gordon, H. S. 1977. "Social Science and Value Judgements." *Canadian Journal of Economics,* 10, 529-46.

Gordon, H. S. 1991. *The History and Philosophy of Social Science.* London: Routledge Publishing.

Gray, J. 1982. "F. A. Hayek and the Rebirth of Classical Liberalism." *Literature of Liberty,* 5, 19-67.

Greenfield, R. L. and Salerno, J. T. 1983. "Another Defense of Methodological A Priorism." *Eastern Economic Journal,* 9, 45-56.

Grunbaum, A. 1976a. "Is Falsifiability the Touchstone of Scientific Rationality? Karl Popper versus Inductivism." In R. S. Cohen et al., eds., *Essays in Memory of Imre Lakatos.* Dordrecht, Holland: D. Reidel, 213-52.

Grunbaum, A. 1976b. "Ad Hoc Auxiliary Hypotheses and Falsificationism." *British Journal for the Philosophy of Science,* 27, 329-62.

Gutting, G., ed. 1980. *Paradigms and Revolutions.* Notre Dame, IN: Notre Dame University Press.

Habermas, J. 1981. "Modernity versus Postmodernity." *New German Critique,* 22, 3-14.

Habermas, J. 1987a. "Philosophy as Stand-in and Interpreter." In K. Baynes, J. Bohman, and T. McCarthy, eds., *After Philosophy.* Cambridge, MA: MIT Press, 296-315.

Habermas, J. 1987b. *The Philosophical Discourse of Modernity.* Cambridge, MA: MIT Press.

Hacking, I. 1979. "Imre Lakatos' Philosophy of Science." *British Journal for the Philosophy of Science,* 30, 381-410.

Hahn, F. H. 1961. "A Stable Adjustment Process for a Competitive Stability. " *Review of Economic Studies,* 39, 62-65.

Hahn, F. H. 1973. "The Winter of Our Discontent." *Economica,* 40, 322-30.

Hahn, F. H. 1983. *Money and Inflation.* Cambridge, MA: MIT Press.

Hahn, F. H. and Negishi, T. 1962. "A Theorem on Non-tatonnement Stability." *Econometrica,* 30, 463-69.

Hamminga, B. 1983. *Neoclassical Theory Structure and Theory Development.* New York: Springer Verlag.

Hamminga, B. 1991. "Comment on Hands." In M. Blaug and N. de Marchi, eds., *Appraising Modern Economics: Studies in the Methodology of Scientific Research Programmes.* Aldershot, England: Edward Elgar, 76-84.

Handler, E. W. 1980a. "The Logical Structure of Modern Neoclassical Static Microeconomic Equilibrium Theory." *Erkenntnis,* 15, 33-53.

Handler, E. W. 1980b. "The Role of Utility and of Statistical Concepts in Empirical Economic Theories: The Empirical Claims of the Systems of Aggregate Market Supply and Demand Functions Approach." *Erkenntnis,* 15, 129-57.

Hands, D. W. 1979. "The Methodology of Economic Research Programmes." *Philosophy of the Social Sciences,* 9, 293-303. (See Chapter 1 of this book.)

Hands, D. W. 1984a. "Blaug's Economic Methodology." *Philosophy of the Social Sciences,* 14, 115-25. (See Chapter 3 of this book.)

Hands, D. W. 1984b. "The Role of Crucial Counterexamples in the Growth of Economic Knowledge: Two Case Studies in the Recent History of Economic Thought." *History of Political Economy,* 16, 59-67. (See Chapter 2 of this book.)

Hands, D. W. 1985a. "Second Thoughts on Lakatos." *History of Political Economy,* 17, 1-16. (See Chapter 4 of this book.)

Hands, D. W. 1985b. "Karl Popper and Economic Methodology: A New Look." *Economics and Philosophy,* 1, 83-99. (See Chapter 6 of this book.)

Hands, D. W. 1985c. "The Structuralist View of Economic Theories: A Review Essay." *Economics and Philosophy,* 1, 303-35.

Hands, D. W. 1988. "Ad Hocness in Economics and the Popperian Tradition." In N. de Marchi, ed., *The Popperian Legacy in Economics.*

Cambridge, England: Cambridge University Press, 121-37. (See Chapter 7 of this book.)

Hands, D. W. 1990a. "Second Thoughts on 'Second Thoughts': Reconsidering the Lakatosian Progress of *The General Theory.*" *Review of Political Economy,* 2, 69-81. (See Chapter 5 of this book.)

Hands, D. W. 1990b. "Thirteen Theses on Progress in Economic Methodology." *Finnish Economic Papers,* 3, 72-76. (See Chapter 10 of this book.)

Hands, D. W. 1990c. "Grunberg and Modigliani, Public Predictions, and the New Classical Macroeconomics." In W. J. Samuels, ed., *Research in the History of Economic Thought and Methodology.* Greenwich, CN: JAI Press, 207-23.

Hands, D. W. 1991a. "Reply to Mäki and Hamminga." In M. Blaug and N. deMarchi, eds., *Appraising Economic Theories: Studies in the Methodology of Scientific Research Programmes.* Aldershot, England: Edward Elgar, 91-102.

Hands, D. W. 1991b. "Popper, the Rationality Principle, and Economic Explanation." In G. K. Shaw, ed., *Economics, Culture and Education: Essays in Honor of Mark Blaug.* Aldershot, England: Edward Elgar, 108-19.

Hands, D. W. 1991c. "The Problem of Excess Content: Economics, Novelty, and a Long Popperian Tale." In M. Blaug and N. de Marchi, eds., *Appraising Economic Theories: Studies in the Methodology of Scientific Research Programs.* Aldershot, England: Edward Elgar, 58-75. (See Chapter 9 of this book.)

Hands, D. W. 1992a. "Falsification, Situational Analysis, and Scientific Research Programs: The Popperian Tradition in Economic Methodology." In N. de Marchi, ed., *Post-Popperian Methodology of Economics.* Boston: Kluwer Publishing. (See Chapter 8 of this book.)

Hands, D. W. 1992b. "Reply to Mark Blaug." In N. de Marchi, ed., *Post-Popperian Methodology of Economics.* Boston: Kluwer Publishing.

Harre, R. 1972. *The Philosophies of Science.* Oxford, England: Oxford University Press.

Hauptli, B. W. 1991. "A Dilemma for Bartley's Pancritical Rationalism." *Philosophy of the Social Sciences,* 21, 86-89.

Hausman, D. M. 1980. "How to Do Philosophy of Economics." In P. D. Asquith and R. Giere, eds., *PSA 1980.* East Lansing, MI: PSA, 353-62.

Hausman, D. M. 1981a. "Are General Equilibrium Theories Explanatory?" In J. C. Pitt, ed., *Philosophy in Economics.* Dordrecht, Holland: D. Reidel, 17-32.

Hausman, D. M. 1981b. *Capital, Profits, and Prices: An Essay in the Philosophy of Economics.* New York: Columbia University Press.

Hausman, D. M. 1985. "Is Falsificationism Unpracticed or Unpracticable?" *Philosophy of the Social Sciences,* 15, 313-19.

Hausman, D. M. 1988. "An Appraisal of Popperian Economic Methodology." In N. de Marchi, ed., *The Popperian Legacy in Economics.* Cambridge, England: Cambridge University Press, 65-85.

Hausman, D. M. 1992. *The Inexact and Separate Science of Economics.* Cambridge, England: Cambridge University Press.

Hayek, F. A. 1967. "The Theory of Complex Phenomena." In *Studies in Philosophy, Politics, and Economics.* Chicago: University of Chicago Press, 22-42.

Hayek, F. A. 1979. *The Counter-revolution in Science.* Indianapolis, IN: Liberty Press.

Hendry, D. 1980. "Econometrics—Alchemy or Science?" *Economica,* 47, 387-406.

Hesse, M. 1988. "Socializing Epistemology." In E. McMullin, ed., *Construction and Constraint.* Notre Dame, IN: University of Notre Dame Press, 97-122.

Hicks, J. R. 1939. *Value and Capital.* Oxford, England: Oxford University Press.

Hicks, J. R. 1976. "Revolutions in Economics." In S. J. Latsis, ed., *Method and Appraisal in Economics.* Cambridge, England: Cambridge University Press, 207-18.

Hirsch, A. and de Marchi, N. 1984. "Methodology: A Comment on Frazer and Boland." *American Economic Review,* 74, 782-88.

Hirsch, A. and de Marchi, N. 1990. *Milton Friedman: Economics in Theory and Practice.* Ann Arbor: University of Michigan Press.

Hollis, M. 1985. "The Emperor's Newest Clothes." *Economics and Philosophy,* 1, 128-33.

Hornsby, J. 1990. "Descartes, Rorty, and the Mind-Body Fiction." In A. Malachowski, ed., *Reading Rorty.* Oxford, England: Basil Blackwell, 41-57.

Hull, D. 1988. *Science as a Process: An Evolutionary Account of the*

Social and Conceptual Development in Science. Chicago: University of Chicago Press.

Hutchison, T. W. 1938. *The Significance and Basic Postulates of Economic Theory.* London: Macmillan (reprint, New York: Augustus M. Kelly, 1960).

Hutchison, T. W. 1960. "Methodological Prescription in Economics: A Reply." *Economica,* 27, 158-60.

Hutchison, T. W. 1976. "On the History and Philosophy of Science and Economics." In S. J. Latsis, ed., *Method and Appraisal in Economics.* Cambridge, England: Cambridge University Press, 181-205.

Hutchison, T. W. 1981. *The Politics and Philosophy of Economics.* New York: New York University Press.

Hutchison, T. W. 1988. "The Case for Falsificationism." In N. de Marchi, ed., *The Popperian Legacy in Economics.* Cambridge, England: Cambridge University Press, 169-81.

Jaffé, W. 1980. "Walras's Economics as Others See It." *Journal of Economic Literature,* 18, 528-49.

Jarvie, I. C. 1972. *Concepts and Society.* London: Routledge and Kegan Paul.

Jarvie, I. C. 1982. "Popper on the Difference between the Natural and Social Sciences." In B. Levinson, ed., *In Pursuit of Truth: Essays in the Philosophy of Karl Popper on the Occasion of His 80th Birthday.* Atlantic Highlands, NJ: Humanities Press, 83-107.

Jarvie, I. C. 1984. "A Plague on Both Your Houses." In J. R. Brown, ed., *Scientific Rationality: The Sociological Turn.* Dordrecht, Holland: D. Reidel, 165-82.

Kaldor, N. 1972. "The Irrelevance of Equilibrium Economics." *Economic Journal,* 82, 1237-55.

Keynes, J. M. 1936. *The General Theory of Employment, Interest, and Money.* Volume 7 of *The Collected Writings of J. M. Keynes.* London: Royal Economic Society, 1973.

Kirzner, I. M. 1976. "On the Method of Austrian Economics." In E. G. Dolan, ed., *The Foundations of Modern Austrian Economics.* Mission, KS: Sheed and Ward, 40-51.

Klamer, A. 1983. *Conversations with Economists.* Totowa, NJ: Rowman and Allenheld.

Klamer, A. 1984. "Levels of Discourse in New Classical Economics." *History of Political Economy,* 16, 263-90.

Klamer, A. 1987. "The Advent of Modernism." Unpublished.

Klamer, A. 1988a. "Economics as Discourse." In N. de Marchi, ed., *The Popperian Legacy in Economics.* Cambridge, England: Cambridge University Press, 259-78.

Klamer, A. 1988b. "Negotiating a New Conversation about Economics." In A. Klamer, D. N. McCloskey, and R. M. Solow, eds., *The Consequences of Economic Rhetoric.* Cambridge, England: Cambridge University Press, 265-79.

Klamer, A. 1990. "The Textbook Presentation of Economic Discourse." In W. J. Samuels, ed., *Economics as Discourse.* Boston: Kluwer Academic, 129-54.

Klamer, A. and McCloskey, D. N. 1988. "Economics in the Human Conversation." In A. Klamer, D. N. McCloskey, and R. M. Solow, eds., *The Consequences of Economic Rhetoric.* Cambridge, England: Cambridge University Press, 3-20.

Klamer, A. and McCloskey, D. N. 1989. "The Rhetoric of Disagreement." *Rethinking Marxism,* 2, 140-61.

Klamer, A., McCloskey, D. N., and Solow, R. M., eds. 1988. *The Consequences of Economic Rhetoric.* Cambridge, England: Cambridge University Press.

Klant, J. J. 1984. *The Rules of the Game.* Cambridge, England: Cambridge University Press.

Klant, J. J. 1988. "The Natural Order." In N. de Marchi, ed., *The Popperian Legacy in Economics.* Cambridge, England: Cambridge University Press, 87-117.

Klappholz, K. and Agassi, J. 1959. "Methodological Prescriptions in Economics." *Economica,* N.S., 26, 60-74.

Knorr-Cetina, K. 1981. *The Manufacture of Knowledge: An Essay on the Constructivist and Contextual Nature of Science.* Oxford, England: Pergamon Press.

Knorr-Cetina, K. 1982. "The Constructivist Programme in the Sociology of Science: Retreats or Advances?" *Social Studies of Science,* 12, 320-24.

Knorr-Cetina, K. 1991. "Epistemic Cultures: Forms of Reason in Science." *History of Political Economy,* 23, 105-22.

Koertge, N. 1971. "Inter-theoretic Criticism and the Growth of Science." In R. C. Buck and R. S. Cohen, eds., *Boston Studies in the Philosophy of Science*, Volume 8. Dordrecht, Holland: D. Reidel, 160-73.

Koertge, N. 1974. "On Popper's Philosophy of Social Science." In K. F. Schaffner and R. S. Cohen, eds., *PSA 1972*. Dordrecht, Holland: D. Reidel, 195-207.

Koertge, N. 1975. "Popper's Metaphysical Research Program for the Human Sciences." *Inquiry*, 19, 437-62.

Koertge, N. 1978. "Toward a New Theory of Scientific Inquiry." In G. Radnitzky and G. Anderson, eds., *Progress and Rationality in Science*. Dordrecht, Holland: D. Reidel, 253-78.

Koertge, N. 1979a. "The Methodological Status of Popper's Rationality Principle." *Theory and Decision*, 10, 83-95.

Koertge, N. 1979b. "The Problems of Appraising Scientific Theories." In P. D. Asquith and H. E. Kyburg, Jr., eds., *Current Research in Philosophy of Science*. East Lansing, MI: PSA, 228-51.

Koertge, N. 1985. "On Explaining Beliefs." *Erkenntnis*, 22, 175-86.

Koppl, R. 1989. "On the Philosophical and Political Vision of Leon Walras." Unpublished.

Kuhn, T. S. 1970a. *The Structure of Scientific Revolutions*. Chicago: University of Chicago Press.

Kuhn, T. S. 1970b. "Logic of Discovery or Psychology of Research?" In I. Lakatos and A. Musgrave, eds., *Criticism and the Growth of Knowledge*. Cambridge, England: Cambridge University Press, 1-12.

Kuhn, T. S. 1970c. "Reflections on My Critics." In I. Lakatos and A. Musgrave, eds., *Criticism and the Growth of Knowledge*. Cambridge, England: Cambridge University Press, 231-78.

Kuhn, T. S. 1971. "Notes on Lakatos." In R. C. Buck and R. S. Cohen, eds., *Boston Studies in the Philosophy of Science*, Volume 8. Dordrecht, Holland: D. Reidel, 137-46.

Kuhn, T. S. 1977. *The Essential Tension*. Chicago: University of Chicago Press.

Kuipers, T. A. F. 1990. "Reduction of Laws and Concepts." *Poznan Studies in the Philosophy of the Sciences and the Humanities*, 16, 241-76.

Kulka, T. 1977. "Some Problems Concerning Rational Reconstructions, Comments on Elkana and Lakatos." *British Journal for the Philosophy of Science*, 28, 325-44.

Lakatos, I. 1968. "Criticism and the Methodology of Scientific Research Programmes." *Proceedings of the Aristotelian Society,* 69, 149-86.

Lakatos, I. 1970. "Falsification and the Methodology of Scientific Research Programmes." In I. Lakatos and A. Musgrave, eds., *Criticism and the Growth of Knowledge.* Cambridge, England: Cambridge University Press, 91-196.

Lakatos, I. 1971a. "History of Science and Its Rational Reconstructions." In R. C. Buck and R. S. Cohen, eds., *Boston Studies in the Philosophy of Science,* Volume 8. Dordrecht, Holland: D. Reidel, 91-136.

Lakatos, I. 1971b. "Replies to Critics." In R. C. Buck and R. S. Cohen, eds., *Boston Studies in the Philosophy of Science,* Volume 8. Dordrecht, Holland: D. Reidel, 174-82.

Lakatos, I. 1974. "The Role of Crucial Experiments in Science." *Studies in the History and Philosophy of Science,* 4, 309-25.

Lakatos, I. 1978a. "Changes in the Problem of Inductive Logic." In J. Worrall and G. Currie, eds., *Mathematics, Science, and Epistemology: Philosophical Papers,* Volume 2. Cambridge, England: Cambridge University Press, 128-200.

Lakatos, I. 1978b. "Anomalies versus 'Crucial Experiments.'" In J. Worrall and G. Currie, eds., *Mathematics, Science, and Epistemology: Philosophical Papers,* Volume 2. Cambridge, England: Cambridge University Press, 211-23.

Lakatos, I. 1978c. "Popper on Demarcation and Induction." In J. Worrall and G. Currie, eds., *The Methodology of Scientific Research Programmes: Philosophical Papers,* Volume 1. Cambridge, England: Cambridge University Press, 139-67.

Lakatos, I. and Musgrave, A., eds. 1970. *Criticism and the Growth of Knowledge.* Cambridge, England: Cambridge University Press.

Langlois, R. N. and Csontos, L. 1989. "Models of Bounded Rationality and the Methodology of Situational Analysis." Paper presented at a conference on "Methodological Problems of Neo-institutional Economics," Uppsala, Sweden, August 1989.

Latour, B. 1990. "Postmodern? No, Simply Amodern! Steps toward an Anthropology of Science." *Studies in the History and Philosophy of Science,* 21, 145-71.

Latour, B. and Woolgar, W. 1986. *Laboratory Life: The Construction of Scientific Facts,* 2nd edition. Princeton, NJ: Princeton University Press.

Latsis, S. J. 1972. "Situational Determinism in Economics." *British Journal for the Philosophy of Science,* 23, 207-45.

Latsis, S. J., ed. 1976a. *Method and Appraisal in Economics.* Cambridge, England: Cambridge University Press.

Latsis, S. J. 1976b. "A Research Programme in Economics." In S. J. Latsis, ed., *Method and Appraisal in Economics.* Cambridge, England: Cambridge University Press, 1-41.

Latsis, S. J. 1983. "The Role and Status of the Rationality Principle in the Social Sciences." In R. S. Cohen and M. W. Wartofsky, eds., *Epistemology, Methodology, and the Social Sciences.* Dordrecht, Holland: D. Reidel, 123-51.

Laudan, L. 1965. "Grunbaum on 'The Duhemian Argument.'" *Philosophy of Science,* 32, 295-99.

Laudan, L. 1977. *Progress and Its Problems.* Berkeley: University of California Press.

Laudan, L. 1979. "Historical Methodologies: An Overview and Manifesto." In P. D. Asquith and H. E. Kyburg, Jr., eds., *Current Research in Philosophy of Science.* East Lansing, MI: PSA, 40-54.

Laudan, L. 1988. "Are All Theories Equally Good? A Dialogue." In R. Nola, ed., *Relativism and Realism in Science.* Boston: Kluwer, 117-34.

Laudan, L. 1990. *Science and Relativism.* Chicago: University of Chicago Press.

Leamer, E. 1983. "Let's Take the Con out of Econometrics." *American Economic Review,* 73, 31-43.

Leijonhufvud, A. 1976. "Schools, 'Revolutions,' and Research Programmes in Economic Theory." In S. J. Latsis, ed., *Method and Appraisal in Economics.* Cambridge, England: Cambridge University Press, 65-108.

Lennon, K. 1990. *Explaining Human Action.* LaSalle, IL: Open Court.

Lipsey, R. G. 1966. *An Introduction to Positive Economics,* 2nd edition. London: Weidenfeld and Nicholson.

Lipsey, R. G. and Steiner, P. O. 1975. *Economics,* 4th edition. New York: Harper and Row.

Livingston, E. 1986. *Ethnomethodological Foundations of Mathematics.* London: Routledge and Kegan Paul.

Lodge, D. 1984. *Small World.* New York: Macmillan.

Lucas, R. E. 1981. "Tobin and Monetarism: A Review Article." *Journal of Economic Literature,* 19, 558-67.

Lynch, M. 1985. *Art and Artifact in Laboratory Science.* London: Routledge and Kegan Paul.

Lynch, M. 1992a. "Extending Wittgenstein: The Pivotal Move from Epistemology to the Sociology of Science." In A. Pickering, ed., *Science as Practice and Culture.* Chicago: University of Chicago Press, 215-65.

Lynch, M. 1992b. "From the 'Will to Theory' to the Discursive Collage: A Reply to Bloor's 'Left and Right Wittgensteinians.'" In A. Pickering, ed., *Science as Practice and Culture.* Chicago: University of Chicago Press, 283-300.

Lyotard, J. F. 1987. "The Postmodern Condition." In K. Baynes, J. Bohman, and T. McCarthy, eds., *After Philosophy.* Cambridge, MA: MIT Press, 73-94.

McCloskey, D. N. 1983. "The Rhetoric of Economics." *Journal of Economic Literature,* 21, 481-517.

McCloskey, D. N. 1984. "Reply to Caldwell and Coats." *Journal of Economic Literature,* 22, 579-80.

McCloskey, D. N. 1985. *The Rhetoric of Economics.* Madison: University of Wisconsin Press.

McCloskey, D. N. 1988a. "Thick and Thin Methodologies in the History of Economic Thought." In N. de Marchi, ed., *The Popperian Legacy in Economics.* Cambridge, England: Cambridge University Press, 245-57.

McCloskey, D. N. 1988b. "The Consequences of Rhetoric." In A. Klamer, D. N. McCloskey, and R. M. Solow, eds., *The Consequences of Economic Rhetoric.* Cambridge, England: Cambridge University Press, 280-93.

McCloskey, D. N. 1988c. "Two Replies and a Dialogue on the Rhetoric of Economics." *Economics and Philosophy,* 4, 150-66.

McCloskey, D. N. 1989a. "Formalism in Economics, Rhetorically Speaking." *Ricerche Economiche,* 43, 57-75.

McCloskey, D. N. 1989b. "Why I Am No Longer a Positivist." *Review of Social Economy,* 47, 225-38.

McMullin, E. 1976. "The Fertility of Theory and the Unit for Appraisal in Science." In R. S. Cohen et al., eds., *Essays in Memory of Imre Lakatos.* Dordrecht, Holland: D. Reidel, 395-432.

McMullin, E. 1979. "The Ambiguity of 'Historicism.'" In P. D. Asquith and H. E. Kyburg, Jr., eds., *Current Research in Philosophy of Science.* East Lansing, MI: PSA, 55-83.

Maddock, R. 1984. "Rational Expectations Macrotheory: A Lakatosian Case Study in Program Adjustment." *History of Political Economy,* 16, 291-309.

Mäki, U. 1986. "Scientific Realism and Austrian Explanation." (Early draft of Mäki, 1990a.)

Mäki, U. 1988a. "How to Combine Rhetoric and Realism in the Methodology of Economics." *Economics and Philosophy,* 4, 89-109.

Mäki, U. 1988b. "Realism, Economics, and Rhetoric." *Economics and Philosophy,* 4, 167-69.

Mäki, U. 1989a. "On the Problem of Realism in Economics." *Ricerche Economiche,* 43, 176-98.

Mäki, U. 1989b. "Mengerian Economics in Realist Perspective." Paper presented at a conference on "Carl Menger and His Legacy in Economics," Duke University, April 1989.

Mäki, U. 1990a. "Scientific Realism and Austrian Explanation." *Review of Political Economy,* 2, 310-44.

Mäki, U. 1990b. "Methodology of Economics: Complaints and Guidelines." *Finnish Economic Papers,* 3, 77-84.

Mäki, U. 1991a. "Comment on Hands." In M. Blaug and N. de Marchi, eds., *Appraising Modern Economics: Studies in the Methodology of Scientific Research Programmes.* Aldershot, England: Edward Elgar, 85-90.

Mäki, U. 1991b. "Persuasion, Plausibility, and Truth: Two Philosophies of the Rhetoric of Economics." In W. Henderson, T. Dudley-Evans, and R. Backhouse, eds., *Economics, Language, and Critical Theory* (forthcoming).

Mäki, U. 1992. "Social Conditioning of Economics." In N. de Marchi, ed., *Post-Popperian Methodology of Economics.* Boston: Kluwer Publishing.

Malachowski, A., ed. 1990. *Reading Rorty.* Oxford, England: Basil Blackwell.

Maxwell, N. 1972. "A Critique of Popper's Views on Scientific Method." *Philosophy of Science,* 39, 131-52.

Merrill, G. H. 1980. "Moderate Historicism and the Empirical Sense of

'Good Science.'" In P. D. Asquith and R. N. Gieve, eds., *PSA 1980,* Volume 1. East Lansing, MI: PSA, 223-35.

Metzler, L. A. 1945. "Stability of Multiple Markets: the Hicks Conditions." *Econometrica,* 13, 277-92.

Milberg, W. 1988. "The Language of Economics: Deconstructing the Neoclassical Text." *Social Concept,* 4, 33-57.

Miller, D. W. 1974. "Popper's Qualitative Theory of Verisimilitude." *British Journal for the Philosophy of Science,* 25, 166-77.

Mirowski, P. 1987. "Shall I Compare Thee to a Minkowski-Ricardo-Leontief-Metzler Matrix of the Mosak-Hicks Type? Or, Rhetoric, Mathematics, and the Nature of Neoclassical Economic Theory." *Economics and Philosophy,* 3, 67-95.

Mirowski, P. 1989. *More Heat than Light: Economics as Social Physics: Physics as Nature's Economics.* Cambridge, England: Cambridge University Press.

Mirowski, P. 1991. "Postmodernism and the Social Theory of Value." *Journal of Post Keynesian Economics,* 13, 565-82.

Mirowski, P. 1992. "Three Vignettes on the State of Economic Rhetoric." In N. de Marchi, ed., *Post-Popperian Methodology of Economics.* Boston: Kluwer Publishing.

Morgan, M. 1988. "Finding a Satisfactory Empirical Model." In N. de Marchi, ed., *The Popperian Legacy in Economics.* Cambridge, England: Cambridge University Press, 199-211.

Morishima, M. 1952. "On the Laws of Change of the Price-Systems in an Economy Which Contains Complementary Commodities." *Osaka Economic Papers,* 7, 101-13.

Morishima, M. 1973. *Marx's Economics.* Cambridge, England: Cambridge University Press.

Morishima, M. and Catephores, G. 1978. *Value, Exploitation, and Growth.* London: McGraw-Hill UK.

Munz, P. 1985. *Our Knowledge of the Growth of Knowledge.* London: Routledge and Kegan Paul.

Munz, P. 1987. "Philosophy and the Mirror of Rorty. " In G. Radnitzky and W. W. Bartley III, eds., *Evolutionary Epistemology, Rationality, and the Sociology of Knowledge.* LaSalle, IL: Open Court, 345-78.

Murphy, N. 1990. "Scientific Realism and Postmodern Philosophy." *British Journal for the Philosophy of Science,* 41, 291-303.

Musgrave, A. 1974. "Logical versus Historical Theories of Confirmation." *British Journal for the Philosophy of Science,* 25, 1-23.

Musgrave, A. 1976. "Method or Madness? Can the Methodology of Research Programmes Be Rescued from Epistemological Anarchism?" In R. S. Cohen et al., eds., *Essays in Memory of Imre Lakatos.* Dordrecht, Holland: D. Reidel, 457-91.

Musgrave, A. 1981. " 'Unreal Assumptions' in Economic Theory." *Kyklos,* 34, 377-87.

Musgrave, A. 1989. "Saving Science from Scepticism." In F. D'Agostina and I. C. Jarvie, eds., *Freedom and Rationality: Essays in Honor of John Watkins.* Dordrecht, Holland: Kluwer Academic, 197-323.

Nagel, E. 1961. *The Structure of Science.* New York: Harcourt, Brace, and World.

Negishi, T. 1958. "A Note on the Stability of an Economy Where All Goods Are Gross Substitutes." *Econometrica,* 26, 445-47.

Nelson, A. 1989. "Human Molecules." Unpublished.

Nelson, A. 1990. "Social Science and the Mental." *Midwest Studies in Philosophy,* 15, 194-209.

Nola, R. 1987. "The Status of Popper's Theory of Scientific Method." *British Journal for the Philosophy of Science,* 38, 441-480.

Nola, R. 1988. "Introduction: Some Issues concerning Relativism and Realism in Science." In R. Nola, ed., *Relativism and Realism in Science.* Boston: Kluwer Academic, 1-35.

Nola, R. 1990. "The Strong Programme for the Sociology of Science, Reflexivity, and Relativism." *Inquiry,* 33, 273-96.

North, D. C. 1990. *Institutions, Institutional Change, and Economic Performance.* Cambridge, England: Cambridge University Press.

Nozick, R. 1977. "On Austrian Methodology." *Synthese,* 36, 353-92.

O'Brien, D. P. 1976. "The Longevity of Adam Smith's Vision: Paradigms, Research Programmes, and Falsifiability in the History of Economic Thought." *Scottish Journal of Political Economy,* 23, 133-51.

Oddie, G. 1986. "The Poverty of the Popperian Program for Truthlikeness." *Philosophy of Science,* 53, 163-78.

Olson, M. 1984. "Beyond Keynesianism and Monetarism." *Economic Inquiry,* 22, 297-322.

Perry, G. L. 1984. "Reflections on Macroeconomics." *American Economic Review,* 74, 401-7.

Petrella, F. 1988. "Henry George and the Classical Scientific Research Program: The Economics of Republican Millennialism." *American Journal of Economics and Sociology,* 47, 239-56.

Pheby, J. 1988. *Methodology and Economics: A Critical Introduction.* London: Macmillan.

Phillips, A. W. 1958. "The Relation between Unemployment and the Rate of Change in Money Wage Rates in the United Kingdom." *Economica,* 282-99.

Pickering, A., ed. 1992. *Science as Practice and Culture.* Chicago: University of Chicago Press.

Popper, K. R. 1961. *The Poverty of Historicism,* 3rd edition. New York: Harper and Row.

Popper, K. R. 1965. *Conjectures and Refutations,* 2nd edition. New York: Harper and Row.

Popper, K. R. 1966. *The Open Society and Its Enemies,* Volume 2, 2nd edition. New York: Harper and Row.

Popper, K. R. 1967. "La Rationalité et le Statut de Principe de Rationalité." In E. M. Classen, ed., *Les Fondements Philosophiques des Systems Economiques.* Paris: Payot, 142-50. (Now translated as Popper, 1985.)

Popper, K. R. 1968. *The Logic of Scientific Discovery,* 2nd edition. New York: Harper and Row (1st English edition, 1959; originally in German, 1934).

Popper, K. R. 1970. "Normal Science and Its Dangers." In I. Lakatos and A. Musgrave, eds., *Criticism and the Growth of Knowledge.* Cambridge, England: Cambridge University Press, 51-58.

Popper, K. R. 1972. *Objective Knowledge.* Oxford, England: Oxford University Press.

Popper, K. R. 1974. "Replies to My Critics." In P. A. Schilpp, ed., *The Philosophy of Karl Popper.* LaSalle, IL: Open Court, 961-1197.

Popper, K. R. 1976a. "The Logic of the Social Sciences." In T. W. Adorno et al., eds., *The Positivist Dispute in German Sociology,* translated by G. Adey and D. Frisby. New York: Harper and Row, 87-104.

Popper, K. R. 1976b. *Unended Quest.* LaSalle, IL: Open Court.

Popper, K. R. 1977. "Part I by Karl R. Popper." In K. R. Popper and J. C. Eccles, eds., *The Self and Its Brain.* New York: Springer-Verlag, 3-223.

Popper, K. R. 1982. *Quantum Theory and the Schism in Physics.* Totowa, NJ: Rowman and Littlefield.

Popper, K. R. 1983. *Realism and the Aim of Science.* Totowa, NJ: Rowman and Littlefield.

Popper, K. R. 1985. "The Rationality Principle." In D. Miller, ed., *Popper Selections.* Princeton, NJ: Princeton University Press, 357-65.

Putnam, H. 1974. "The 'Corroboration' of Theories." In P. A. Schilpp, ed., *The Philosophy of Karl Popper.* LaSalle, IL: Open Court.

Quinn, P. 1972. "Methodological Appraisal and Heuristic Advice." *Studies in the History and Philosophy of Science,* 3, 135-49.

Quirk, J. and Saposnik, R. 1968. *Introduction to General Equilibrium Theory and Welfare Economics.* New York: McGraw-Hill.

Radnitzky, G. 1982. "Knowing and Guessing: If All Knowledge is Conjectural, Can We Then Speak of Cognitive Progress? On Persistent Misreadings of Popper's Work." *Zeitschrift fur Allgemeine Wissenschaftstheorie,* 13, 110-21.

Radnitzky, G. 1986. "Towards an 'Economic' Theory of Methodology." *Methodology and Science,* 19, 124-47.

Radnitzky, G. 1989. "Falsificationism Looked at from an 'Economic' Point of View." In K. Gavroglu, Y. Goudaroulis, and P. Nicolacopoulos, eds., *Imre Lakatos and Theories of Scientific Change.* Boston: Kluwer Academic, 383-95.

Radnitzky, G. 1991. "Refined Falsificationism Meets the Challenge from the Relativist Philosophy of Science." *British Journal for the Philosophy of Science,* 42, 273-84.

Radnitzky, G. and Bartley, W. W. III, eds. 1987. *Evolutionary Epistemology, Rationality, and the Sociology of Knowledge.* LaSalle, IL: Open Court.

Rappaport, S. 1988a. "Economic Methodology: Rhetoric or Epistemology?" *Economics and Philosophy,* 4, 110-28.

Rappaport, S. 1988b. "Arguments, Truth, and Economic Methodology." *Economics and Philosophy,* 4, 170-72.

Redman, D. A. 1991. *Economics and the Philosophy of Science.* New York: Oxford University Press.

Remenyi, J. V. 1979. "Core Demi-core Interaction: Towards a General Theory of Disciplinary and Subdisciplinary Growth." *History of Political Economy,* 11, 30-63.

Rescher, N. 1990. *A Useful Inheritance: Evolutionary Aspects of the Theory of Knowledge.* Savage, MD: Roman and Littlefield.

Rizzo, M. J. 1982. "Mises and Lakatos: A Reformulation of Austrian Methodology." In I. M. Kirzner, ed., *Method, Process, and Austrian Economics.* Lexington, MA: Lexington Books, 53-72.

Robbins, L. 1979. "On Latsis's *Method and Appraisal in Economics:* A Review Essay." *Journal of Economic Literature,* 17, 996-1004.

Rorty, R. 1965. "Mind-Body Identity, Privacy, and Categories." *Review of Metaphysics,* 19, 24-54.

Rorty, R. 1970. "In Defense of Eliminative Materialism." *Review of Metaphysics,* 24, 112-21.

Rorty, R. 1979. *Philosophy and the Mirror of Nature.* Princeton, NJ: Princeton University Press.

Rorty, R. 1982a. *Consequences of Pragmatism.* Minneapolis: University of Minnesota Press.

Rorty, R. 1982b. "Contemporary Philosophy of Mind." *Synthese,* 53, 323-48.

Rorty, R. 1988. "Is Natural Science a Natural Kind?" In E. McMullin, ed., *Construction and Constraint.* Notre Dame, IN: University of Notre Dame Press, 49-74.

Rorty, R. 1989a. "Philosophy as Science, as Metaphor, and as Politics." In A. Cohen and M. Dascal, eds., *The Institution of Philosophy.* La-Salle, IL: Open Court, 13-33.

Rorty, R. 1989b. *Contingency, Irony, and Solidarity.* Cambridge, England: Cambridge University Press.

Rosenau, P. M. 1992. *Post-modernism and the Social Sciences: Insights, Inroads, and Intrusions.* Princeton, NJ: Princeton University Press.

Rosenberg, A. 1980. "A Skeptical History of Microeconomic Theory." *Theory and Decision,* 12, 79-93.

Rosenberg, A. 1981. *Sociobiology and the Pre-emption of Social Science.* Baltimore: Johns Hopkins University Press.

Rosenberg, A. 1983. "If Economics Isn't a Science, What Is it?" *Philosophical Forum,* 14, 296-314.

Rosenberg, A. 1985. "Methodology, Theory, and the Philosophy of Science." *Pacific Philosophical Quarterly,* 66, 377-93.

Rosenberg, A. 1986. "Lakatosian Consolations for Economists." *Economics and Philosophy,* 2, 127-39.

Rosenberg, A. 1988a. *Philosophy of Social Science*. Boulder, CO: Westview Press.

Rosenberg, A. 1988b. "Economics Is Too Important to Be Left to the Rhetoricians." *Economics and Philosophy*, 4, 129-49.

Rosenberg, A. 1988c. "Rhetoric Is Not Important Enough for Economists to Bother About." *Economics and Philosophy*, 4, 173-75.

Rosenberg, A. 1989. "Superseding Explanation versus Understanding: The View from Rorty." *Social Research*, 56, 479-510.

Rossetti, J. 1990. "Deconstructing Robert Lucas." In W. J. Samuels, ed., *Economics as Discourse*. Boston: Kluwer Academic, 225-43.

Rothbard, M. N. 1976. "Praxeology: The Methodology of Austrian Economics." In E. G. Dolan, ed., *The Foundations of Modern Austrian Economics*. Mission, KS: Sheed and Ward, 19-39.

Ruccio, D. F. 1991. "Postmodernism and Economics." *Journal of Post Keynesian Economics*, 13, 495-510.

Salanti, A. 1987. "Falsification and Fallibilism as Epistemic Foundations of Economics: A Critical View." *Kyklos*, 40, 368-92.

Salmon, W. C. 1966. *The Foundations of Scientific Inference*. Pittsburgh: University of Pittsburgh Press.

Samuelson, P. A. 1941. "The Stability of Equilibrium: Comparative Statics and Dynamics." *Econometrica*, 9, 97-120.

Samuelson, P. A. 1942. "The Stability of Equilibrium: Linear and Nonlinear Systems." *Econometrica*, 10, 1-25.

Samuelson, P. A. 1944. "The Relation between Hicksian Stability and True Dynamic Stability." *Econometrica*, 12, 256-57.

Sargent, T. and Wallace, N. 1976. "Rational Expectations and the Theory of Economic Policy." *Journal of Monetary Economics*, 2.

Scarf, H. 1960. "Some Examples of Global Instability at the Competitive Equilibrium." *International Economic Review*, 1, 147-72.

Schmidt, R. H. 1982. "Methodology and Finance." *Theory and Decision*, 14, 391-413.

Schumpeter, J. A. 1954. *History of Economic Analysis*. New York: Oxford University Press.

Schwartz, J. 1991. "Reduction, Elimination, and the Mental." *Philosophy of Science*, 58, 203-20.

Searle, J. R. 1983. *Intentionality: An Essay in the Theory of Mind.* New York: Cambridge University Press.

Searle, J. R. 1991. "Intentionalistic Explanations in the Social Sciences." *Philosophy of the Social Sciences,* 21, 332-44.

Shaw, G. K. 1988. *Keynesian Economics: The Permanent Revolution.* Aldershot, England: Edward Elgar.

Simon, H. A. 1976. "From Substantive to Procedural Rationality." In S. J. Latsis, ed., *Method and Appraisal in Economics.* Cambridge, England: Cambridge University Press, 129-48.

Smith, A. 1776. *An Inquiry into the Nature and Causes of the Wealth of Nations.* New York: Random House, Modern Library edition, 1937.

Steedman, I. 1975. "Positive Profits with Negative Surplus Value." *Economic Journal,* 74, 114-23.

Stegmuller, W. 1978. "A Combined Approach to the Dynamics of Theories." *Theory and Decision,* 9, 29-75.

Stegmuller, W., Balzer, W., and Spohn, W., eds. 1982. *Philosophy of Economics.* Berlin: Springer-Verlag.

Stich, S. 1983. *From Folk Psychology to Cognitive Science.* Cambridge, MA: MIT Press.

Stigler, G. 1965. "The Development of Utility Theory." In *Essays in the History of Economics.* Chicago: University of Chicago Press, 66-155.

Suppe, F. 1977. *The Structure of Scientific Theories,* 2nd edition. Urbana: University of Illinois Press.

Suppe, F. 1979. "Theory Structure." In P. D. Asquith and H. E. Kyburg, Jr., eds., *Current Research in Philosophy of Science.* East Lansing, MI: PSA, 317-38.

Suppes, P. 1979. "The Role of Formal Methods in the Philosophy of Science." In P. D. Asquith and H. E. Kyburg, Jr., eds., *Current Research in Philosophy of Science.* East Lansing, MI: PSA, 16-27.

Susser, B. 1989. "The Sociology of Knowledge and Its Enemies." *Inquiry,* 32, 245-60.

Takayama, A. 1972. *International Trade.* New York: Holt, Reinhart, and Winston.

Tarascio, V. J. 1971. "Value Judgements and Economic Science." *Journal of Economic Issues,* 5, 98-102.

Tichy, P. 1974. "On Popper's Definitions of Verisimilitude." *British Journal for the Philosophy of Science,* 25, 155-60.

Tobin, J. 1980. *Asset Accumulation and Economic Activity.* Chicago: University of Chicago Press.

von Mises, L. 1949. *Human Action: A Treatise on Economics.* New Haven, CN: Yale University Press.

von Mises, L. 1978. *The Ultimate Foundation of Economic Science,* 2nd edition. Kansas City, MO: Sheed Andrews and McMeel.

Walras, L. 1954. *Elements of Pure Economics,* translated by W. Jaffe from the 4th definitive edition, 1926. Homewood, IL: Irwin.

Warnke, G. 1985. "Hermeneutics and the Social Sciences: A Gadamerian Critique of Rorty." *Inquiry,* 28, 339-57.

Watkins, J. 1958. "Confirmable and Influential Metaphysics." *Mind,* 67, 344-65.

Watkins, J. 1969. "Comprehensively Critical Rationalism." *Philosophy,* 44, 57-62.

Watkins, J. 1970. "Imperfect Rationality." In R. Berger and F. Cioffi, eds., *Explanation in the Behavioral Sciences.* Cambridge, England: Cambridge University Press, 167-217.

Watkins, J. 1971. "CCR: A Refutation." *Philosophy,* 46, 56-61.

Watkins, J. 1975. "Metaphysics and the Advancement of Science." *British Journal for the Philosophy of Science,* 25, 91-121.

Watkins, J. 1978. "The Popperian Approach to Scientific Knowledge." In G. Radnitzky and G. Anderson, eds., *Progress and Rationality in Science.* Dordrecht, Holland: D. Reidel, 23-43.

Watkins, J. 1984. *Science and Scepticism.* Princeton, NJ: Princeton University Press.

Watkins, J. 1987. "Comprehensively Critical Rationalism: A Retrospect." In G. Radnitzky and W. W. Bartley III, eds., *Evolutionary Epistemology, Rationality, and the Sociology of Knowledge.* LaSalle, IL: Open Court.

Weintraub, E. R. 1979. *Microfoundations.* Cambridge, England: Cambridge University Press.

Weintraub, E. R. 1982. Review of *The Mathematical Experience* by P. J. Davis and R. Hersh. *Journal of Economic Literature,* 20, 114-15.

Weintraub, E. R. 1983. "On the Existence of a Competitive Equilibrium, 1930-1954." *Journal of Economic Literature,* 21, 1-39.

Weintraub, E. R. 1985a. "Appraising General Equilibrium Analysis." *Economics and Philosophy,* 1, 23-37.

Weintraub, E. R. 1985b. *General Equilibrium Analysis: Studies in Appraisal.* Cambridge, England: Cambridge University Press.

Weintraub, E. R. 1988a. "The NeoWalrasian Program is Empirically Progressive." In N. de Marchi, ed., *The Popperian Legacy in Economics.* Cambridge, England: Cambridge University Press, 213-27.

Weintraub, E. R. 1988b. "On the Brittleness of the Orange Equilibrium." In A. Klamer, D. N. McCloskey, and R. M. Solow, eds., *The Consequences of Economic Rhetoric.* Cambridge, England: Cambridge University Press.

Weintraub, E. R. 1991. *Stabilizing Dynamics: Constructing Economic Knowledge.* Cambridge, England: Cambridge University Press.

Wible, J. R. 1992. "Cost-Benefit Analysis, Utility Theory, and Economic Aspects of Peirce's and Popper's Conceptions of Science" (forthcoming).

Wisdom, J. O. 1963. "The Refutability of 'Irrefutable' Laws." *British Journal for the Philosophy of Science,* 13, 303-06.

Wisdom, J. O. 1970. "Situational Individualism and Emergent Group Properties." In R. Borger and F. Cioffi, eds., *Explanation in the Behavioral Sciences.* Cambridge, England: Cambridge University Press, 271-96.

Wisdom, J. O. 1987. *Challengeability in Modern Science.* Aldershot, England: Gower Publishing.

Wong, S. 1978. *The Foundation of Paul Samuelson's Revealed Preference Theory.* London: Routledge and Kegan Paul.

Worrall, J. 1978. "The Ways in Which the Methodology of Scientific Research Programmes Improves on Popper's Methodology." In G. Radnitzky and G. Anderson, eds., *Progress and Rationality in Science.* Dordrecht, Holland: D. Reidel, 45-70.

Worrall, J. 1982. "Scientific Realism and Scientific Change." *Philosophical Quarterly,* 32, 201-31.

Worrall, J. 1989a. "Structural Realism: The Best of Both Worlds?" *Dialectica,* 43, 99-124.

Worrall, J. 1989b. "Why Both Popper and Watkins Fail to Solve the Problem of Induction." In *Freedom and Rationality: Essays in Honor of John Watkins.* Dordrecht, Holland: Kluwer Academic, 257-96.

Zahar, E. G. 1973. "Why Did Einstein's Programme Supersede Lorentz's?" *British Journal for the Philosophy of Science,* 24, 95-123, 223-62.

Zahar, E. G. 1983. "The Popper-Lakatos Controversy in the Light of 'Die Beidn Grundprobleme der Erkenntnistheorie.'" *British Journal for the Philosophy of Science,* 34, 149-71.

Zahar, E. G. 1989. "John Watkins on the Empirical Basis and the Corroboration of Scientific Theories." In *Freedom and Rationality: Essays in Honor of John Watkins.* Dordrecht, Holland: Kluwer Academic, 325-41.

Index